AN INVESTIGATION OF 3 VIRUS DISEASES OF *VANILLA FRAGRANS* (SALISB) AMES IN THE KINGDOM OF TONGA

By SEMISI PULE PONE

THE AUTHOR

This photo was taken on 10 May, 1985 on graduation with a Bachelor of Science from the University of Auckland, New Zealand.

The name on my passport at the time was Semisi Pule also known as Semisi Pule Pone although I used Semisi Pone for short.

This Thesis reproduction is dedicated to the
memory of my parents **Mr Samisoni Manu Sisitoutai
Ponepate Pule** (locally known in Tonga as Samisoni Pone) and **Mrs
Tupou Fele'ave Ponepate Pule** (locally known in Tonga as Tupou
Pone) of the Kingdom of Tonga, who were so proud of my academic
achievements.

ACKNOWLEDGEMENTS

I wish to acknowledge the help and support of the following people;

My supervisor, Dr M.N.Pearson, for the discussions and various other ways in which he made immense contribution to this study. I am also grateful for the supply of antisera to CyMV and ORSV. My thanks to his wife Amira, for welcoming us in her home.

Dr D. H. Stechmann, former Manager of the Tonga-German Plant Protection Project, for the faith and trust that got me this scholarship.

Mr Nga Tama for his expertise with immunisation of rabbits.

Mr John Cole for his help with compost medium and plants in the glasshouse.

John White and Terry for giving me a helping hand with electron microscopy when it was sorely needed.

I am also grateful to all other staff and students of the Botany Department who deserve some mention, but cannot fit into this short space, for the help they had generously given.

Special thanks goes to Mr Konrad Englberger, Manager of the Tonga-German Plant Protection Project (Deutsche Gesellschaft fuer Technische Zusammenarbeit (GTZ), and other staff for their support during this study. The staff of the Research Division and MAFF

Headquarters also deserve some thanks for all the help I had received during the field work in Tonga and the supply of some information for Chapter 1. Many thanks to Mr 'Etuate Pāngata'a, my field and laboratory assistant, whose help during the field surveys and laboratory was invaluable.

The financial assistance given by the German Agency for Overseas Student Exchange (Deutscher Akademicher Austauschdienst (DAAD), with support from GTZ, made this study possible. I therefore wish to express my thanks to the staff members at DAAD headquarters and the staff of the Embassy of the Federal Republic of Germany, in Wellington, for administering my scholarship so well.

Last, but not least, I wish to thank my wife Peata and Semisi junior for keeping me going.

ABSTRACT.

Three viruses found to infect *Vanilla fragrans* in Tonga were identified as Cymbidium Mosaic Virus (CyMV), Odontoglossum Ringspot Virus (ORSV) and a 'potyvirus' (VPV). CyMV and ORSV are two well known virus diseases of Orchids, but the 'potyvirus' appear to be an unknown virus. Characterization of the 'potyvirus' (VPV), in this study, showed that it is a definite member of the Potato Virus Y group and may be an uncited potyvirus.

VPV was partially purified from frozen VPV infected Vanilla tissue from Tonga. An antiserum was developed to VPV by immunizing a rabbit with partially purified VPV preparation and post absorption of antibodies to immunogenic Vanilla plant protein that contaminated the partially purified preparation. The gamma globulin was purified from the blood serum and part of it was conjugated with an enzyme (alkaline phosphatase). Purified gamma globulin to CyMV and ORSV were also conjugated with the same enzyme.

The Double Antibody Sandwich Enzyme Linked Immuno-Sorbent Assay (DAS-ELISA) was optimized by testing for the optimum concentrations of coating and enzyme conjugate to be used. Optimum coating concentrations for all 3 antisera was 1µl/ml and the enzyme conjugate dilution of 1:1600 was found to be optimal. A serial dilution of a partially purified virus of known concentration was used to test the sensitivity of DAS-ELISA which was found to be about 85 nanograms per millilitre.

The relative effects of the 3 viruses on Vanilla plants were assessed using young Vanilla shoots from newly planted cuttings. Vanilla plants were inoculated with the 3 viruses by grafting and by mechanical inoculation. VPV was found to cause severe stunting, diffuse chlorotic streaks and terminal necrosis in newly infected plants. The leaves of new shoots that develop from the same infected plant exhibit symptoms of slight leaf distortion and localized chlorotic streaks which may turn necrotic. CyMV and ORSV did not show any observable effect on the growth of the Vanilla plants nor did they induce any severe symptoms on the Vanilla foliage during the pathogenicity study.

Field surveys based on the presence of the symptoms attributed to VPV found this virus to be widespread on the islands of Vava'u and Tongatapu with incidences of symptomatic plants ranging fro 0-83% per plantation and averaging about 12%.

Surveys of young plantations of less than 2 years old found 100% correlation between symptomatic cuttings and the development of symptoms on the new shoots. It is concluded that the agent for long distance spread of VPV are Vanilla plant cuttings.

Three Vanilla plots were mapped and the two dimensional and linear distribution of VPV symptomatic plants in the three plots were analyzed for any significant deviation from the Poisson distribution. Two of the plots (Plots 1 and 3) had highly significant deviations (P=0.025) from the Poisson distribution in a two dimensional analysis. The third plot (Plot 2) did not differ significantly from the Poisson distribution at P=0.05. Since potyviruses are mostly transmitted by aphids, and VPV, can be transmitted by aphids. It was concluded that aphids have a possible role in the transmission of VPV inside Plots 1 and 3.

The linear distribution of VPV symptomatic plants, along Vanilla rows, was found to deviate significantly (P=0.05) from the Poisson distribution in 3 rows in Plot 1. This demonstrates the possible effect of transmitting the virus along Vanilla rows during cultural practices.

....TABLE OF CONTENTS....

....CHAPTER 1....

INTRODUCTION TO *VANILLA FRAGRANS*

....CHAPTER 2 ....

DISEASES OF *VANILLA FRAGRANS*

....CHAPTER 3....

CHARACTERIZATION OF THE VANILLA POTYVIRUS

3.1   INTRODUCTION

3.2   MATERIALS AND METHODS

3.3   RESULTS

3.4 DISCUSSION

....CHAPTER 4....

DEVELOPING A DETECTION METHOD FOR VPV, CyMV AND ORSV.

4.1 INTRODUCTION

4.2   MATERIALS AND METHODS

4.3 RESULTS

4.4 DISCUSSION

....CHAPTER 6....

EPIDEMIOLOGY OF VPV

6.1   INTRODUCTION

6.2   MATERIALS AND METHODS

6.3   RESULTS

....CHAPTER 7....

GENERAL DISCUSSIONS AND CONCLUSIONS...136

LIST OF TABLES

LIST OF FIGURES

LIST OF PLATES

LIST OF APPENDICES

MAP OF THE SOUTH PACIFIC.

NIUA FO'OU

NIUA TOPUTAPU

MAP OF THE KINGDOM OF TONGA.

VAVA'U
GROUP

PACIFIC OCEAN

HA'APAI
GROUP

TONGATAPU
GROUP

0   50   100 MILES

O'ATA

....CHAPTER 1....

Introduction to *Vanilla fragrans* (Salisb.) Ames syn. *Vanilla planifolia* (Andrews).

1.1 Origin and History

Correl (1953) recorded *Vanilla fragrans* as indigenous to Mexico and some parts of Central America where it grew wild as a climber in the tropical rainforest. The flowers of *V. fragrans* are pollinated by bees native to Mexico. When the beans were ripe, the ancient Aztecs collected and cured them for flavoring a drink referred to as 'chocolatl', based on ground cocoa seeds. The Aztecs called the plant 'Tlilxochitl' meaning black pod, which is the appearance of the beans after being cured (see section 1.4).

The Spanish introduced the recipe of 'chocolatl' to Europe, in the late 16th century, where factories were set up to manufacture 'chocolate'. It is believed that this was the beginning of the Vanilla trade, and of chocolate. The word Vanilla comes from the Spanish 'vainilla' or 'vaynilla' which mean 'little pod'. The Spanish later changed to cinnamon flavored chocolate, but Vanilla remained the favorite in France. The French were later responsible for the distribution and cultivation of *V. fragrans* in their many tropical colonies.

1

V. fragrans was introduced, in the form of cuttings, to ReUnion (1793), Java (1819) and Mauritius (1827), but cultivation for commercial purposes was not possible without natural pollinators. An artificial method of pollination, discovered by a former slave in ReUnion Islands in 1841, made systematic commercial cultivation possible. *V. fragrans* was then introduced to Madagascar (Malagasy Republic) (1840), Tahiti (1848), Seychelles (1866) and the Comoros Islands in 1893. Correll (1953) also listed Guadeloupe, Martinique, Puerto Rico, Dominica and various other countries where *V. fragrans* was introduced. According to Tiollier (1980 unpubl.), *V. fragrans* was introduced to Tonga by French visitors from the ReUnion Islands in the 19th century. It was planted at Vava'u and left to grow wild until cultivation started in the 1950s. (see section 1.3)

Paul van der Grijp's paper, 'Brown Gold. Official expectations and local assessment of the Tongan Vanilla production', on the Journal de la Société des Océanistes, 104, 1997 page 93-103 describe the state of the Vanilla Industry in Tonga and how it developed in Vava'u. The problems of the Vanilla export groups and the apparent success of the Tongan Vanilla internationally.

He states that it was a French adventurer, Charles Simonet, who came to Tonga on the French naval vessel Astrolabe, in 1827, that introduced the Vanilla to Tonga. Charles Simonet left the ship at Tongatapu and later settled at Holeva Village, Vava'u. He sent his 14 year old son, Soane Patita, to school in Tahiti and the boy returned several years later with a Tahitian wife and some Vanilla cuttings which were planted at Vava'u and left to grow wild there. Vanilla was, again, reintroduced in the 1950's as a 'Tongan Government Project' initiated by Crown Prince Tupouto'a Tungi (later King Taufa'ahau Tupou IV) when he was Minister of Agriculture in 1953.

Figure 1.1 Distribution of the Genus *Vanilla* (Bouriquet, 1954)

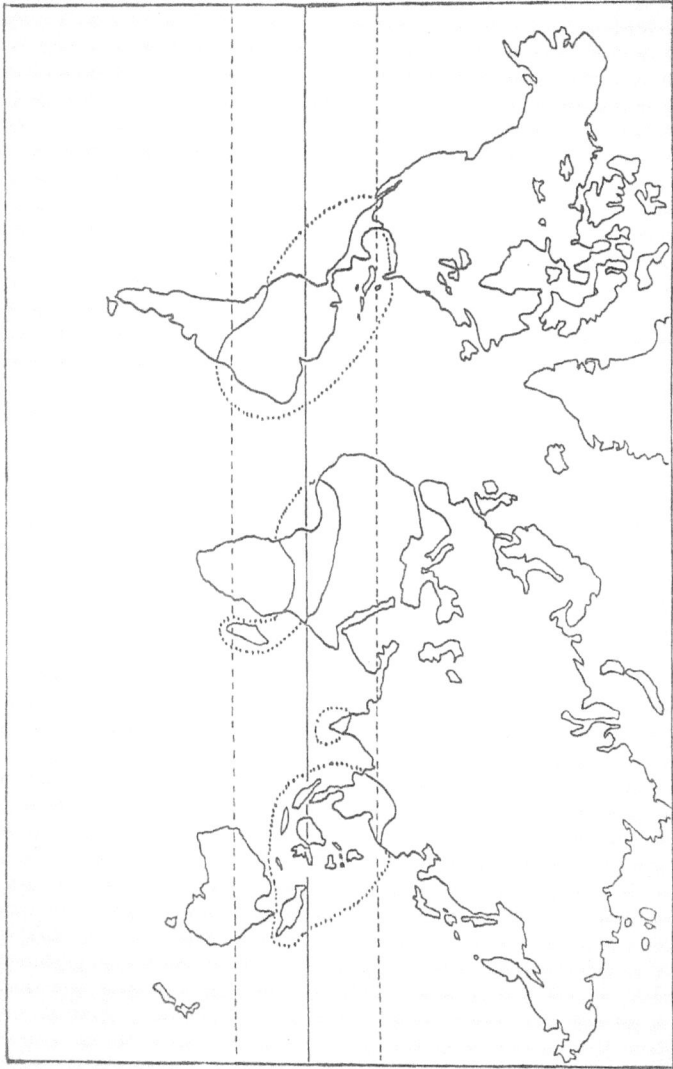

Figure 1.1 Distribution of the genus Vanilla (Bouriquet, 1954).

1.2 Botany

1.2.1 Taxonomy and Cytology

Vanilla is a genus of the family *Orchidaceae*. Porteres, in Bouriquet (1954), described 110 species distributed throughout the tropics (Figure 1.1). The basic chromosome number of the genus is x=16, but several species like *V. fragrans*, *V. pompona* Schiede and *V. tahitiensis* J.W.Moore are diploids with 2n=32 (Purseglove *et al*, 1981). Heim (1954), quoted by Martin (1963), also reported *V. phaeantha* Reich, *V. dilloniana* Correl, *V. barbellata* Reich and a *V. planifolia* x *V. phaentha* f1 hybrid as diploids. According to Tonnier (1951), quoted by Martin (1963), a tetraploid with 64 chromosomes was found and named *V. haapape*.

1.2.2 Description (Plate 1)

Vanilla fragrans has a monopodial, cylindrical stem, 1-2 cm in diameter, with widely spaced, fleshy, alternate, subsessile oblong-elliptic to lanceolate leaves, of a dark green colour, measuring 8-20 cm long and 4-8 cm wide, with indefinite terminal growth of up to 15-20 meters long (Richter, 1965; Purseglove et al, 1981).

Roots are produced singly, opposite the leaves, and are about 2-4 mm in diameter. They are whitish in colour, adventitious, aerial and cling to trees, supports or substrate. Roots that reach the ground branch out in the top 2-10 cm of the soil where the humus layer provide a rich nutrient source for optimum growth. A *Rhizoctonia* mycorrhiza has often been associated with Vanilla roots (Bernard, 1909 and Burges, 1936 as quoted by Leakey, 1970; Alconero, 1969a, b).

4

Plate 1. *Vanilla fragrans* **vines trained on support tree (***Jatropha curcas***)**

This is one of the Vanilla research plots at the Vaini Research Station, Tongatapu, Kingdom of Tonga established by the French Vanilla and Spice Project which I took over as part of my Vanilla virus disease and management research projects (1989-1991).

The flowers are produced in terminal racemes, with up to 20-30 flowers, each measuring about 10 cm in diameter (Plate 2).

Purseglove *et al* (1981) describe the flowers as large, waxy, fragrant, pale greenish-yellow, fugacious, with three oblong lanceolate sepals, obtuse to subacute, 4-7 cm long and 11.5 cm wide. The pedicel is very short with inferior, cylindrical tricarpillary ovary 4-7 cm long and 3-5 cm in diameter.

Plate 2. *Vanilla fragrans* flowers

There are three petals with the two upper petals resembling the sepals in shape , but slightly smaller, while the lower petal is modified as a trumpet-shaped labellum or lip, which is shorter than the other perianth lobes and is 4-5 cm long and 1.5-3 cm broad at its widest point. It is attached to the column which it envelops. The column is 3-5cm long and is attached to the labellum for most of its length. It is hairy in its inner surface, bearing at its tip the single stamen containing the two pollen masses and below is the concave sticky stigma, which is separated from the stamen by the thin flaplike rostellum.

Plate 3. *Vanilla fragrans* **beans**

1.2.3 Ecology

Vanilla fragrans thrives best in a hot, moist, insular climate. The optimum temperature is 20-30°C, with an average of around 27°C. An evenly distributed rainfall of 2,000-2,500 mm per year, with two drier months to check vegetative growth and bring the vines to flower, is considered ideal. *V. fragrans* can grow in any soil provided it is not too heavy and waterlogged. The best approach is a well drained soil with a thick layer of mulch (Tiollier, 1980. unpublished.).

1.3 Cultivation

Three species of Vanilla are cultivated commercially, *V. fragrans*, *V. pompona* and *V. tahitiensis*. *Vanilla pompona* is cultivated mainly in the French West Indies (Dominica, Martinique and Guadeloupe), *V. tahitiensis* is mainly cultivated in French Polynesia. The major vanilla of commerce is *V. fragrans*, which is now mostly cultivated in Madagascar, Comoros Is, ReUnion Is., Indonesia and Tonga (Table 1.1).

In *worldatlas.com*, Economic reports, it shows these countries as the major producers of cured vanilla beans in 2016. Compare with figures in Table 1.1.	
Country	Production/tonnes/cured beans
1. Madagascar	2926
2. Indonesia	2304
3. China	885
4. Mexico	513
5. Papua New Guinea	502
6. Turkey	303
7. Tonga	180
8. French Polynesia	24
9. ReUnion	21
10. Malawi	20
11. Comoros Is.	15
12. Guadeloupe	11
13. Zimbabwe	11

Table 1.1 World Exports and Imports of Cured Vanilla Beans (tonnes)					
EXPORTS	**1982**	**1983**	**1984**	**1985**	**1986**
1. Comoros Is.	259	177	100	181	-
2. Malagasy Republic	1060	1099	827	710	-
3. ReUnion Is.	11	9	14	12	-
4. Indonesia	115	234	154	190	-
5. Tonga *	12	9	14	12	20
IMPORTS					
1. Australia	5	13	10	7	7
2. Canada	30	29	31	31	28
3. United Kingdom	12	20	20	9	14
4. Belgium - Luxembourg	15	14	11	5	10
5. Denmark	21	17	21	29	33
6. France	301	261	202	154	199
7. West Germany	182	200	175	137	190
8. Italy	11	14	21	14	4
9. Netherlands	20	31	23	30	20
10. Spain	-	13	13	6	6
11. Switzerland	32	29	41	35	43
12. United States	884	977	841	743	1001
13. Japan	19	42	36	40	60

Source - Fruit and Tropical Products, Commonwealth Secretariat, London, December 1987. * - Project Performance Audit Report, Tonga Development Bank, May 1988, unpublished.

1.3.1 Cultivation of *V. Fragrans* in Tonga.
(Adapted from Victor Tiollier, 1980 unpublished).

Vanilla fragrans is grown from a cutting, at the base of a support tree, *Jatropha curcas*, with a spacing of 1.5 m between plants and 2.5 m between rows giving rise to approximately 2200 - 2500 plants per hectare. The vanilla vine is trained on the support tree to facilitate easy reach when working on the plant (see Plate 1). The process of training vines is called 'looping'. It involves bringing down long shoots, breaking off 1-2 leaves and burying that part of the vine in the mulch where it will root.

The support tree also provides shade for the vanilla plant which requires 30-50% shading during vegetative growth. The trees are pruned and shaped to maintain a good shade all year round. They are also pruned a month or so before the shoots are 'topped' to initiate flowering. This activity should coincide with a dry period, where vines are exposed to the sun to help with flowering. Cuttings take 3-4 years to flower and may produce beans for 12 years or more.

Flowers are hand pollinated in the morning, as they open, because they last for only one day. An experienced worked can pollinate 1,000 to 1,500 flowers in one morning. Beans take from 6-8 months to mature.

1.4 Curing

V. fragrans beans (see Plate 3) split at the distal end when they become ripe. This splitting reduces the value of the beans, when sold as whole beans. Curing is an artificial method of controlled ripening where the mature distal end turns greenish-yellow. Beans are harvested and cured before they split. It also improves the quality of the beans whether they are for the whole bean market or for extracts.

Curing involves killing the harvested, mature, green beans by dipping in hot water at 63°C for 3 minutes. The water is quickly drained and the beans are 'sweated' in closed wooden boxes, lined with blankets, for at least 24 hours. The temperature, not only kills the beans, but also initiates enzymatic reactions which produce vanillin from a glucoside. Vanillin is the main component of the more than 100 volatile constituents that give natural Vanilla its fuller flavor and aromatic fragrance. Artificial Vanilla does not contain any other aromatic compounds beside vanillin also known as vannaline (Purseglove *et al*, 1981).

The beans are taken out for sun drying during the sunny hours of the day and are returned to the sweating blankets overnight. This process takes from 3-7 days until the beans become supple. To prevent over-drying, the beans are moved to the shade to slow down the drying process. Shade drying may last from 1-2 months depending on when curer thinks the beans are ready. This means the beans are supple and roll between the fore finger and thumb when pressed. They should have lost about 60% of their water content at this stage.

The final stage is storing of the beans in blanket lined trunks. As the enzymatic reactions that produce vanillin, and its associated compounds, are hydrolytic reactions, the beans have to be taken out of the trunks and shade dried over night, every 10 days, to remove the water and to prevent them from going mouldy. After about 2 months the curing is complete and white crystals of vanillin can sometimes be seen on the surface of the cured beans. Beans that are cured properly can be stored for many years without any problems.

(Adapted from Dequaire, 1979 unpublished).

This curing method is known as the Bourbon Method (Paul van der Grijp's paper, 'Brown Gold. Official expectations and local assessment of the Tongan Vanilla production', on the Journal de la Société des Océanistes, 104, 1997, page 93-103) . Other methods have also been used for curing vanilla beans including mechanised curing in Madagascar where large scale curing is carried out (Theodose, 1972).

1.5 World Production and Markets for cured vanilla beans.

The world market for Vanilla is presently limited, with a maximum world consumption between 1,500-2,000 tonnes annually, compared with an increasing output and use of vannaline, a synthetic substitute, which is currently at 12,000 tonnes valued at $US 120 million a year. Production of natural Vanilla varies widely from year to year with the Malagasy Republic, and the ReUnion Islands in the Indian Ocean, supplying the bulk of the global natural Vanilla consumption. Although natural Vanilla seem to be losing to synthetic vannaline, in the world market, the First International Congress on Vanilla , in ReUnion in September, 1987, agreed that as natural Vanilla is an important export earner, producers in the Indian Ocean should take measures to protect natural Vanilla from synthetic substitutes (Fruit and Tropical Products, Commonwealth Secretariat, London, December 1987). Therefore, it is hoped that smaller producer countries like Tonga will benefit from the measures to protect Vanilla taken by the larger producers in the Indian Ocean.

The largest producer of cured Vanilla beans is the Malagasy Republic. The Comoros Is, Indonesia, ReUnion Is and Tonga also produce cured Vanilla beans on a smaller scale. The biggest importer of cured Vanilla beans is the United States followed by France and West Germany. Other countries also import Vanilla to a smaller extent (see Table 1.1).

1.6 The Tongan Vanilla Industry

Cured Vanilla bean is the third largest agricultural export from the Kingdom of Tonga, in terms of foreign currency earnings. The bigger foreign currency earners are currently bananas and coconuts products.
During 1981-1984 cured Vanilla bean export was earning more than bananas and appear to have greater potential. The crop has higher value per unit volume, requires less work and less capital inputs. The product is also more durable and easier to transport. For a comparison, a tonne of green bananas is worth about $TOP 300 whereas a tonne of cured

Vanilla bean is worth $TOP 95,000 at 1987 figures. One Tongan pa'anga is equivalent to one Australian dollar.

Table 1.2 Area under cultivation in the Tonga Islands			
Year	In production (ha)	Not in production (ha)	Total (ha)
1979	22	151	173
1980	40	247	287
1981	No information	N.I.	322
1982	90	230	320
1983	145	335	580
1984	145	335	580
1985	323	468	791
1986	317	461	778
1987	477	403	880

Source - Project Performance Audit Report, Tonga Development Bank, May 1988, unpublished.

Note: A random sample of 150 *V. fragrans* plots from Vava'u and 63 from Tongatapu were selected from the Vanilla Project list of growers to be surveyed for the presence of virus symptoms in 1987. It was found that 80% of the plots selected from Vava'u and 63.5% from Tongatapu had been neglected or abandoned. Therefore, the area under effective Vanilla cultivation may be much less than the official figures quoted above.

Table 1.3 Cured Vanilla Beans production in Tonga between 1979-1987									
Year	1979	1980	1981	1982	1983	1984	1985	1986	1987
Metric tonnes	2	5	9	12	9	14	12	20	19

Source - Project Performance Audit Report, Tonga Development Bank, May 1988, unpublished.

1.6.1 Area under *Vanilla fragrans* cultivation.

Vanilla is grown entirely by small holder growers with a maximum land area of about 3.5 hectares, with a few large scale growers. Most Vanilla growers grow less than one hectare of Vanilla. The Vanilla is grown on the growers own land together with other subsistence crops like taro, yams, cassava and vegetables. Vanilla is grown mostly in Vava'u but cultivation of Vanilla is increasing in popularity in the other islands, especially Tongatapu (see Table 1.2).

1.6.2 Production of cured Vanilla beans in Tonga.

The production of cured Vanilla beans has been increasing steadily in the past 10 years although there is some variation. A reduction in production in some years have been due to droughts and hurricanes. For example, Hurricane Isaac in 1982 appear to have affected the production of 1983 (see Table 1.3).

1.6.3 The Vanilla Development Project

This project was initiated in January 1984 with the objective of establishing a new 300 hectares of Vanilla and to rehabilitate an existing 250 hectares. Additional objectives include provision of administration services, provision of curing sheds, recruiting of a Curing Specialist, provision of Advisory Centers and a Credit Line to Vanilla Growers through the Tonga Development Bank. The project was to be implemented over 4 years completing in June, 1987 (Fa'anunu, 1985, unpublished). The effect of this project can probably be seen in the rapid increase in the area used to grow Vanilla between 1984 and 1987 (See Table 1.2).

1.6.4 Curing and Export

Some Vanilla Growers cure their own Vanilla Beans while others sell their produce, as green beans, to other growers or private curers. These growers and curers export their own produce under private licence, but only on a small scale. By far, the most important curers and exporters are the Tonga Commodities Board (TCB) and the Friendly Islands Marketing Co-operative (FIMCO). They cure and export the bulk of the total Vanilla Production in Tonga.

1.7 Diseases of *Vanilla fragrans*.

There are a substantial number of reports on diseases of *Vanilla fragrans*. These will be covered in Chapter 2.

....CHAPTER 2....

DISEASES OF *VANILLA FRAGRANS.*

2.1 Comment on the diseases of *Vanilla fragrans.*

A large number of pathogens have been described to cause diseases of *V. fragrans.* However, only two of these pathogens seem to cause serious diseases of *V. fragrans* worldwide.

Glomerella vanillae (Zimm.) Petl. (synonymous with *Calospora vanillae* Mass., Bouriquet, 1934; 1954) causing anthracnose (see Section 2.3) and *Fusarium oxysporum* Schlecht ex Fr formae specialis *vanillae* (Tucker) Gordon (synonymous with *Fusarium batatatis* Woll. var. Vanillae Tucker, 1927; Leakey, 1970) causing root rot (see Section 2.10).

Other diseases reported to attack Vanilla are discussed in the latter sections.

The virus diseases of Vanilla are relatively a new era of investigation and are treated in a separate section (Section 2.14). So far reports concerning the virus diseases of Vanilla seem to be restricted to the countries in the Pacific.

2.2 Algal Diseases

Bouriquet (1954) recorded *Cephaleuros virescens* Kunze and *C. henningsii* Schmidle to cause algal leaf spots of Vanilla. *Phycopeltis*

16

aurea Karst was also reported by Dingley *et al* (1981) to cause an algal leaf Spot of Vanilla in Tonga.

2.3 Anthracnose

Anthracnose has been reported to be caused by *Glomerella vanillae* and its conidial state *Colletotrichum vanillae* Scalia but other *Colletotrichum sp*, in association with a *Gloeosporium sp*, have been isolated from diseased Vanilla.

Correl (1953) reported that *Calospora vanillae* causes a serious disease of Vanilla in the Mascarene, Comoros and Seychelles Islands, the West Indies, Tahiti and Colombia. The vegetative apex of the vine, the leaves, aerial roots and beans are attacked. Lionnet (1958) reported *Glomerella vanillae* as the only serious disease of Vanilla in the Seychelles, being responsible for the disappearance of Vanilla Plantations from some of the islands. It is thought that this disease attacks Vanilla only under certain conditions of humidity and vine debility.

Plenet (1958) described *G. vanillae* as the 'Black Disease' of Vanilla in the ReUnion Islands and appears to be causing decline in Vanilla cultivation. *G. vanillae* has also been reported to be the fungus most associated with soft rot of Vanilla in Ceylon (Sri Lanka) (Petch and Ragunathan, 1927). *G. vanillae* also attacks Vanilla in Madagascar, but can be controlled through reducing excessive shade and improving ventilation inside the plantation (Bouriquet 1932, Dequaire 1979, unpublished). Removal and burning of diseased vines and planting new plantations on virgin land were also recommended (Correll 1953; Lionnet 1958).

A *Colletotrichum sp.* and a *Gloeosporium sp.* have also been reported to attack *V. fragrans* in Ceylon (Sri Lanka), Ragunathan (1924), Petch and Ragunathan (1927). It is not known whether this *Colletotrichum sp.* is the conidial state of *Glomerella vanillae* as reported in Correl (1953), Lionnet (1958) and Plenet (1958). Ragunathan (1924) inoculated

17

detached leaves, stem and living plants in the laboratory with these fungi. All attempts to induce infection through unwounded surfaces failed. When inoculations were made on to wounded plucked leaves or wounded living vines, the conidia germinated and formed appressoria but failed to make further progress. However, all inoculations on stem cuttings in moist dishes were successful.

Ciferri (1926) listed 36 species of *Gloeosporium spp.* and *Colletotrichum spp.* reported to be associated with anthracnose of Orchids including Vanilla. A study of the pathogenicity of three strains of *Gloeosporium* and two of *Colletotrichum* isolated from five different Orchids, including Vanilla was made. A series of inoculation tests that were made with these strains on a large range of Orchids, including Vanilla, resulted in 281 failures to produce infection, 8 successes, and eleven doubtful infections. In a second series of inoculations, in which the leaves were coated with collodion and castor oil, a high percentage of successes were obtained. It was concluded that the fungi involved are only weak secondary parasites.

2.4 Bacterial Diseases

Bacterial diseases reported to attack Vanilla include *Phytomonas* (*Pseudomonas*) *cattleyae* which attack Orchids including *V. fragrans* (Ark and Thomas, 1947) and *Phytomonas* (*Pseudomonas*) *solanacearum* which caused heavy losses in *V. fragrans* plantations in the French Antilles (Stehle, 1955). *Bacterium solanacearum*, the causal organism of the Bacterial Wilt of Castor Beans (*Ricinus communis* L.), was reported to cause a wilt in *V. fragrans* plants after being inoculated with a delicate needle prick (Smith and Godfrey 1922).

2.5 Blast and Mildew

Cibes and Childers (1949) referred to a disease of *V. fragrans* caused by *Phytophthora jatrophae* as blast but the same disease was referred to as mildew by Stehle (1954). *P. jatrophae* Jens, was reported from Madagascar (Bouriquet, 1934) to cause rot of Vanilla pods. The pods

18

becomes less turgid and longitudinally wrinkled, dropping off easily. It was also reported to cause the 'Black Disease' of Vanilla in the ReUnion Islands in association with *Calospora vanillae* (Plenet, 1958).

Various other species of *Phytophthora* have also been reported to attack Vanilla in various countries. *Phytophthora palmivora* (Butl.) Butl. has been reported from ReUnion Islands (Kopp 1930) and French Polynesia (Tsao 1985, unpublished). In ReUnion, *P. palmivora* was associated with another soil fungus in causing a disease that destroyed entire plantations of Vanilla. Disease symptoms include the appearance of black lesions on stem, leaves and pods causing them to rot. In French Polynesia, *P.palmivora* together with *P. nicotianae* van Breda de Haan var. p*arasitica* (Dastur) Waterhouse was reported to attack *Vanilla tahitiensis*.

However, Vanilla in Madagascar was reported to be resistant to *P. palmivora* (Report of the Phytopathology Division, Madagascar, 1963). Improved cultivation practices was recommended as the control of the disease (Maublanc and Barat, 1928).

2.6 Brown Spot of *Vanilla fragrans*.

Nectria vanillae Zimm. (synonym *N. vanillicola* Henn., *N.tjibodensis* Penzi and Sacc.) has been named as the cause of brown spotting on *V. fragrans* (Bouriquet, 1954).

2.7 Nematodes

Stier (1984, unpublished) reported a nematode, *Pratylenchus brachyurus* Godfrey, to attack *V. fragrans* under experimental conditions in Tonga. It is not known whether it is a problem in the field.

2.8 Nutrient Deficiency

Cibes and Childers (1946) reported the effect of the calcium, nitrogen, phosphorus and potassium deficiencies on *V. fragrans*. Calcium deficiency did not show any obvious effect on Vanilla plants. Potassium deficient plants appear darker green with a smaller girth. The most obvious effects on the Vanilla plants were nitrogen and potassium deficiencies. Plants that lacked nitrogen turned straw yellow with very limited growth, but root development was not affected. Potassium deficient plants resemble Vanilla plants attacked by *Fusarium* root rot in the field. The decayed and rotted root system of these plants is the obvious reason for the resemblance.

2.9 Post-harvest diseases

A number of fungal moulds have been reported to infest cured Vanilla beans in storage and in transit. *Aspergillus niger*, *Pencillium lividum*, *P. vanillae* and *P. rugulosum* were found on cured Vanilla beans exported from Madagascar to the United States. These various fungi were mainly associated with the attachment point of the beans to the peduncle on beans that were poor in vanillin content. This problem is associated with beans that were harvested before they mature. A treatment of 1% vanillin   is usually sufficient to control these fungi (Bouriquet, 1943). Moreau and Moreau (1962) found cured Vanilla pods from Madagascar to be contaminated by *A. niger* and *Penicillium charlesii* upon their arrival in France. Infection appears to have originated at the point of attachment. The lack of aeration was thought to have caused the fungal growth.

2.10 Root Rot

Fusarium oxysporum f.sp. *vanillae* causes root rot of Vanilla (Tucker, 1924, 1925, 1927; Alconero, 1968). It has been reported to be the most limiting factor to Vanilla production in Puerto Rico (Tucker, 1927; Correll, 1953), Dominica (Dept. Agri. Report, 1946; 1949), Uganda (Leakey, 1970; Ashley, 1980) and Guatemala (Fiester and Elliot, 1966). It has also been reported to be widespread in Madagascar (Bouriquet, 1959; Correll, 1953). The fungus attacks both ground and aerial roots of Vanilla when they are in infected soil. The root portion attacked turns dark brown and decays. Proliferation of new aerial roots will occur, but again these become infected as they reach the soil. The older vines wilt and die off as their capacity to produce more roots is exhausted. The length of time it takes for the vine to die off depend on the size and vigor of the plants and severity of the infection. Some diseased plants may live for several years. New root infections have been closely associated with mechanical damage. A sample of 48 root systems studied histologically found 98% to be closely associated with mechanical damage caused by insects, nematodes and other agents (Alconero, 1968). An inspection of 12 Vanilla plantings in Puerto Rico, covering 100 acres (40 hectares) showed that plantations which were 6 months old or less were free from attack by a species of *Fusarium*, presumably *F.oxysporum* f.sp. *vanillae*, while plantations that were one year old or more were attacked. The disease attacked and destroyed the steele and cortex of the roots. Disease incidence ranged from 1-90% (Jensen, 1940).

A mycorrhizal fungus on Vanilla, *Rhizoctonia solani* Kuhn, has been reported to become parasitic on Vanilla roots at certain times (Alconero, 1969a,b). It was also reported that the mycorrhizal infections by *R. solani*, of mature roots, may be a predisposing factor for the parasitic infection of Vanilla roots by *F. oxysporum* f.sp. *vanillae* (Alconero and Santiago, 1969).

Breeding of a Vanilla hybrid resistant to root rot caused by *F. oxysporum f. sp. vanillae* has been attempted as a control strategy. *V. phaeantha*, which is resistant to *Fusarium* root rot and *V. fragrans* were cross pollinated and the resulting hybrids were grown and observed for any signs of resistance. Twenty six plants, out of forty, grew vigorously in *Fusarium* infested soil, but it took them 8 years, instead of 3-4 years, to flower and the fruit dropped before they matured. It was concluded that a back crossing to *V. fragrans* was necessary (Theis and Jimenez, 1957). Dequaire (1979, unpubl.) stated that the breeding of a Vanilla variety resistant to *Fusarium* root rot has been carried out in Madagascar for 30 years with no success. It was suggested that good cultural management can control this disease in Madagascar.

2.11 Rust

Uredo scabies Cooke (synonym *Uromyces joffrini* Del.) was reported to cause rust of *V. fragrans* (Bouriquet, 1954).

2.12 Miscellaneous Diseases.

Turconi (1924) recorded another disease to have attacked the young beans of *Vanilla fragrans* at the Botanical gardens, Pavia, in Italy. The disease appeared on the young beans as a brownish discoloration, on the proximal end, and progressed downwards. Fruiting bodies of *Botrytis cinerea* finally appeared and covered the whole surface. It was thought that excessively damp conditions enabled *B. cinerea* to invade and attack the tissue of the young pods. Cifferi (1928) isolated a similar fungus which was assumed to be the same fungus isolated by Turconi. The fungus was found to be pathogenic on Vanilla leaves and beans. A species of *Botrytis cinerea*, from another host, was tested negative. Pathogenicity of the two fungi on other hosts were markedly different and therefore it was concluded that the fungus isolated from Vanilla must be a different species. Hence, Ciferri named it *Sclerotinium vanillae*.

Another fungal disease reported to attack Vanilla was *Pyrenochaete vanillae* (Verwoerd and Du Pleiss, 1934).

2.13 Diseases of *V. fragrans* reported from Tonga

Despite the increasingly large area used to grow Vanilla, there has been no conclusive work done on diseases of Vanilla in Tonga. A survey conducted in 1973 on supposed root rot of Vanilla caused by *Fusarium oxysporum Schlect.*, on Vava'u, concluded that 22-50% of plants in 100 vanilla plantings surveyed were attacked by this fungus (Daft and Merrick, 1973 unpubl.). However, no description of the symptoms was given, or the fungus isolated from the roots, nor any pathogenicity work reported.

Dingley *et al* (1981), in a survey of agricultural pests and disease in the Pacific Islands reported the following fungi to have been isolated from *V. fragrans.*

1. *Chaetothrium sp.* (sooty blotch)
2. *Colletotrichum sp.* (leaf dieback)
3. *Fusarium oxysporum f.sp. vanillae* (Tucker) Gordon (root rot)
4. *Glomerella cingulata* (Stonem.) Spauld & Schrenk (anthracnose)
(*G.cingulata* is the sexual phase of *Colletotrichum gloeosporioides* (Penzig) Penzig and Sacc., normally associated with dieback and anthracnose disease of many crops in Tonga)
5. *Nectria haematococca* Berk. & Br. (basal root rot of vine)
6. *Cephaleuros sp.* (algal leaf spot)
7. *Phycopelta aurea* Karst. (algal leaf spot).

Corticium rolfsii Curzi (perfect state *Athelia rolfsii* (Curzi) C.C. Tu & Kimbr., have also been observed to attack Vanilla vines in the field, at Vava'u, especially during the wet season. The use of dead green mulch appear to aggravate this problem.

Although these reports have drawn attention to the presence of these potential pathogens, it is not known whether they are widespread diseases or isolated cases of no significance.

2.14 Virus Diseases

The first report of suspected viruses on *Vanilla fragrans*, in Tonga, was made by van Velsen (1979). A mottle was described on Vanilla leaves thought to be caused by a virus but subsequent examination of the Vanilla leaf sap under the electron microscope found no virus particles (van Velsen, pers. comm.) Another report of viruses on Vanilla was made by Mossop and Fry (1984), who found flexuous filamentous virus particles in Vanilla leaf sap, from Rarotonga, showing symptoms of mosaic and leaf distortion. It was suggested that the particles were of a potyvirus.

However the first detailed report of viruses on Vanilla was made on work done on *Vanilla tahitiensis* in French Polynesia. Three viruses, Cymbidium Mosaic Virus (CyMV), Odontoglossum Ringspot Virus (ORSV) and a potyvirus were found. The incidence of the CyMV and ORSV were quite low (<1%) in the plantations surveyed. The potyvirus was found to be more common with 9% of the plants surveyed infected (Wisler *et al*, 1987a, 1987b).

The reports from French Polynesia prompted inspection of Vanilla in Tonga. Some plants exhibiting symptoms of sunken chlorotic streaks, distortion and necrotic spots on leaves, and extensive necrosis on the stems, were found. Four healthy and five diseased samples of Vanilla leaves and stem, showing the symptoms described, were sent to the University of Florida for viral examination. The subsequent findings on the five diseased samples reported the presence of flexuous, filamentous virus particles were observed under the electron microscope using negatively stained leaf dips. Cytoplasmic inclusions similar to those of the potyvirus group were observed under the light microscope. Serological tests (SDS Immunodiffusion) was negative to CyMV (2.14.2) and ORSV (2.14.3). No evidence of a virus presence was detected in the healthy samples (Zettler, 1986; letter to the Vanilla Agronomist, MAFF, Tonga).

Investigations carried out in Tonga during 1986 found no fungal growth on Potato Dextrose Agar (PDA), from tissue isolated from within necrotic Vanilla stem, leaves and beans. However, various fungi which are not known to cause disease on Vanilla, were found to grow on PDA from tissue of the surface lesions. It was concluded that these fungi are secondary parasites rather than the primary cause of the necrosis (Pone and Nai, 1985 unpubl.).

A report from G.C.Wisler, University of Florida, Gainsville (letter to Vanilla Agronomist, MAFF, Tonga, 1986) also confirm the absence of fungi from the necrotic tissue of virus infected Vanilla plants.

2.14.1 Identification and preliminary survey of viruses in Tonga.

A survey was planned and conducted on two Vanilla plantings on the northern island of Vava'u and one on the main island of Tongatapu in late 1986 (Pearson and Pone, 1988). A total of 30 samples were collected per plantation, ten each from plants exhibiting symptoms of (A) sunken chlorotic streaks on young leaves, with brown scablike lesions (<0.5cm) and necrotic lesions (<2cm) on older leaves and vines, and severe distortion of the usually smooth leaf margin (B) leaf mottle (C) no symptoms.

These samples were tested for virus presence using DAS-ELISA (Chapter 4) and electron microscopy of negatively stained leaf dips. DAS-ELISA was used to test for CyMV and ORSV using antisera to the New Zealand isolates of these viruses from Cymbidium. All the samples were examined for potyviruslike particles as well as CyMV and ORSV under the electron microscope. Results are presented in Table 2.1. The Vanilla Potyvirs will be referred to as VPV.

2.14.2 Cymbidium Mosaic Virus (CyMV, Plate 4)

Table 2.1 Percentage incidence of viruses in leaf samples of *V. fragrans* from Tonga (30 samples).				
Symptom type	VPV	CyMV	ORSV	Not detected
A	67	63	27	13
B	3	50	20	40
C	0	23	3	77

CyMV has flexuous filamentous particles about 475 nanometers (nm) long and 13 nm wide (Francki, 1970). It was described from Cymbidium orchids by Jensen (1951). CyMV causes mosaic and necrosis in several genera of orchids (Jensen 1951; Jensen and Gold, 1955; Kado and Jensen, 1964) and can reduce growth of *Cymbidium* orchids significantly (Pearson and Cole, 1986). It is a fairly stable virus which can survive in crude sap for up to one week at room temperature and up to 70°C for 10 minutes. It multiplies in *Cymbidium* orchid leaves to very high concentrations, yielding about 360mg per kilogram of fresh leaves. It contain 6% RNA and is a member of the Potato Virus X group. It has no known vector but can be easily transmitted mechanically. Diagnostic hosts are *Chenopodium amaranticolor*, *Cassia occidentalis*, *C. tora* and *C. bicapsularis*. Assay species include *C. amaranticolor*, *Datura stramonium* and *C. occidentalis* (Francki, 1970).

Plate 4. Cymbidium Mosaic Virus (CyMV).
Bar = 200 cm. This sample was a negatively stained leaf dip from a Cymbidium orchid in the University of Auckland glasshouse. Note the large amount of virus particles in the picture.

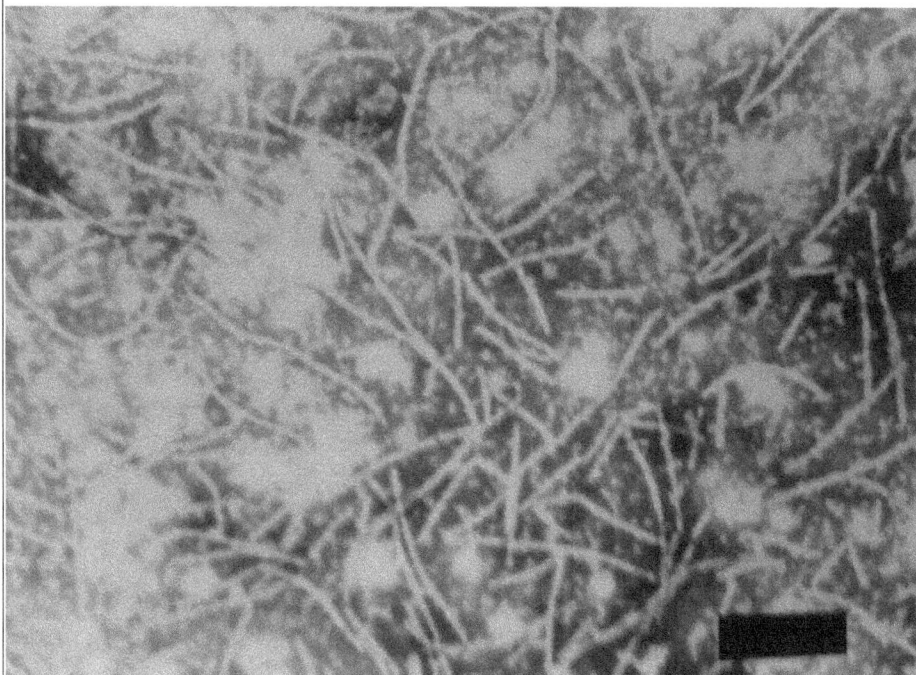

Plate 5.   Odontoglossum Ringspot Virus (ORSV). This sample was also a negatively stained leaf dip from the University of Auckland glasshouse. Bar = 200 nm

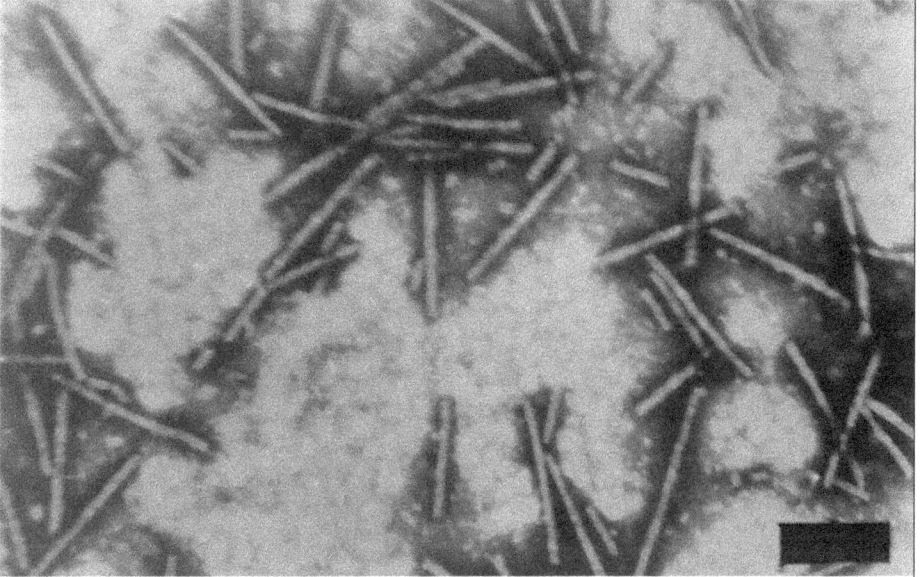

2.14.3 Odontoglossum Ringspot Virus (ORSV, Plate 5).

ORSV has rod shaped particles about 300nm long and 18nm wide infecting a wide range of orchids systemically with no apparent symptoms (Paul, 1975), but causes ringspot symptoms of *Odontoglossum grande* (Jensen and Gold, 1951) and hence the name. ORSV can cause significant reduction in growth of *Cymbidium* orchids (Pearson and Cole, 1986). It is very stable surviving in orchid sap, at room temperature, for over 10 years and up to 90°C for 10 minutes (Edwardson and Zettler, 1986). It multiplies to very high concentrations in *Nicotiana tabacum cv samsum*,yielding up to 1 gram virus per kilogram of leaf tissue. ORSV particles contain about 5% RNA and is a member of the Tobamo Virus Group. No vector is known for this virus, but it is very contagious mechanically. Diagnostic hosts include *Nicotiana tabacum cv Xanthi-nc.* Assay species are *C. quinoa* and *N. tabacum cv Xanthi -nc* (Paul, 1975).

2.14.4 Ecology of CyMV and ORSV.

CyMV and ORSV are both distributed worldwide in cultivated orchids (Francki, 1970; Paul, 1975). They are regarded to be the most prevalent viruses of cultivated orchids in the world (Zettler *et al*, 1978; Wisler *et al*, 1979; Wisler *et al*, 1982). Although they have no efficient vectors, the ease with which they can be mechanically transmitted seem to help spread these viruses under glasshouse conditions whereas orchids in the wild appear to be uninfected. A survey carried out at various locations in Florida and Guatemala found that cultivated orchids had 36% of 298 samples infected wnereas 150 samples collected from the wild orchids were non-infected (Zettler *et al*, 1978). The results of this study highlights the role of routine cultural practices in spreading these two viruses. Therefore, it is highly likely that these two viruses are spread in the same manner in Vanilla plantations.

2.14.5 Potyviruses in Orchids

Several potyviruses have been reported to infect Orchids. Leseman and Koenig (1985) found a strain of Bean Yellow Mosaic Virus (BYMV) to infect 20 species of *Masdevallia* exported from the United States to West Germany. Infected plants exhibited symptoms of chlorotic streak mosaic on young leaves and surface irregularities on older leaves. Some infected plants did not show symptoms at all. The virus particle length was measured to be 745 nanometers (nm) long and parts of pinwheels were found in leaf thin sections. The virus induced severe necrosis on inoculated *Nicotiana clevelandii* and local lesions on *Chenopodium quinoa*.

Another virus was found in *Orchis militaris*, exhibiting symptoms of leaf mosaic and growth reduction. Particle length was 745 nm and parts of pinwheels were also found in leaf thin sections. The virus causes a systemic infection of *N. clevelandii* and local lesions on *C. quinoa*. Immuno Electron Microscopy found the virus to react strongly with antiserum to Turnip Mosaic Virus (TuMV) but only weakly with other antisera. Further testing using direct and indirect ELISA found the virus to be slightly different from TuMV (Leseman and Vetten, 1985).

An unidentified potyvirus measuring 774 nm and also inducing pinwheel formation was found in *Cypripedium calceolus*. The symptoms exhibited by infected plants were chlorotic and later necrotic streaks as well as leaf deformation. The virus has not been transmitted experimentally yet. All these potyviruses are regarded as potentially dangerous to Orchids (Leseman and Vetten, 1985).

2.14.16 The Research Programme

This study first characterized and identified the potyvirus (VPV), which was later renamed Vanilla Necrosis Potyvirus (VNPV) (Chapter 3) isolated from Vanilla samples from Tonga. CyMV and ORSV have been described by Francki (1970) and Paul (1975) respectively, but VPV appears to be a previously undescribed virus. The Double Antibody Sandwich - Enzyme Linked Immunosorbent Assay (DAS-ELISA) was chosen as the main detection method for the three viruses because of its sensitivity and capacity to test large numbers of plant material. Antisera were already available to CyMV and ORSV, therefore an antiserum was produced to VPV.

The antisera to CyMV, ORSV and VPV were conjugated with an enzyme and the DAS-ELISA procedure was optimized using these antisera (Chapter 4). The relative effects of the 3 viruses on *Vanilla fragrans* (Chapter 5) and the epidemiology of VPV (Chapter 6) were also investigated. Implications of the results with respect to virus disease control are discussed in Chapter 7.

....CHAPTER 3....

CHARACTERISATION OF THE VANILLA POTYVIRUS (VPV).

3.1 Introduction

3.1.1 General comments on virus characterisation.

For 'new' viruses, a description of its properties may be required, therefore an investigation of its characteristics is essential. When an unknown virus disease is being investigated, a preliminary characterisation often starts with the determination of particle morphology. This is especially useful for rod shaped and filamentous viruses, which can be identified as belonging to a group from particle size only. Other characteristics are then determined. An appropriate diagnostic test is usually developed to identify the virus after the first preliminary characterisation (Hamilton *et al*, 1981).

Differentiation between viruses is based on their different characteristics. These characteristics enable the grouping of viruses with similar characters together. A knowledge of relationships between viruses and the group to which they belong can help in predicting virus behavior (Hill, 1984), and therefore become invaluable in the field control of these viruses. For example, the knowledge that a virus is a potyvirus enables the prediction of non-persistent transmission by aphids and therefore field control techniques can be oriented towards aphid control.

The Potyvirus group derived its name from its type member Potato Virus Y. It is the largest and economically the most important of the 25 plant virus groups recognised by the International Committee for the Taxonomy of Viruses (Matthews, 1979). Members of the Potyvirus Group have flexuous filamentous particles 680 - 900 nm long and 11 nm wide, with helical symmetry and pitch of 3.4 nm. They have a single polypeptide coat protein with a molecular weight between 32,000 to 36,000 units. Their nucleic acid is a linear positive sense single stranded RNA and they induce characteristic cylindrical or conical cytoplasmic inclusions, appearing as pinwheels when seen in transverse sections. Potyviruses are transmitted experimentally by mechanical inoculations and also in a non-persistent manner by aphids (Matthews, 1982).

The ability to assay viruses is an essential requirement for the study, and identification, of viruses. Four general methods of assay are used;

(i) Infectivity (qualitative assay based on local lesions and quantitative assay based on number of individuals infected),
(ii) Physical (spectrophotometry)
(iii) Serological (Microprecipitation test, Immuno-osmophoresis, Immuno-Sorbent Electron Microscopy, Enzyme Linked Immuno-Sorbent Assay),
(iv) Chemical (dry weight determination, complementary DNA probes).

Infectivity tests are the most basic since it is only through biological activity that we know we are dealing with a virus (Matthews, 1981). Local lesion assays are semi quantitative but various factors like the age of the plant, genetic variation of the host, position of the leaf on the plant, nutrition of the plant, water supply, temperature, light intensity, season of the year and time of day all influence the number of lesions produced by the plant. The use of the 'half-leaf' method have been used to avoid any bias by the above effects but is wasteful of plants and hence a Latin Square Design, where appropriate, is more effective (Matthews, 1981).

3.1.2 Host Range

Host range refers to the range of hosts infected by the virus investigated. It is usually characteristic for some viruses and has been used as a guide to virus identification, but it is now of limited importance (Hamilton *et al*, 1981). Most Potyviruses have a restricted host range. In general, potyviruses found in monocots tend to have a more restricted host range than those found in dicots. The most susceptible host species to potyviruses are *Chenopodium quinoa, C. amaranticolor* and *Nicotiana clevelandii*. These plants have been experimentally infected with 26, 22 and 23 potyviruses, respectively. The most susceptible families are Aizoaceae, Amaranthaceae and Chenopodiaceae (Hollings and Brunt, 1981b).

A second more important objective is to discover suitable plants for maintenance of the virus for laboratory work (Ross, 1964). The usefulness of host plants in the laboratory depends on distribution of the virus in the host, whether it is localised or systemic. Systemic hosts can be used for multiplying the virus quickly for purification purposes, especially if purification from the original host is difficult. The local lesion hosts are used for infectivity assays as well as thermal inactivation point (TIP), dilution end point (DEP) and longevity in vitro (LIV) tests. A third important use of the knowledge of the wider range of a virus's hosts is in its application in the field, where alternative hosts of the virus infecting a crop are removed from the immediate vicinity of the crops. This can be especially helpful with the control of aphid transmitted viruses.

3.1.3 Physical Properties

For many years, Plant Virologists used the host range (3.1.2), thermal inactivation point, dilution end point and the longevity *in vitro* to obtain an indication of the stability of a virus and its concentration (Walkey, 1985). They are now not considered of any great diagnostic value (Hamilton *et al*, 1981), but they are quite valuable in giving an indication of the stability and concentration of a virus in sap which may be helpful in developing purification procedures. Ross (1964) suggest it should be generally realized that the data generated vary and do not necessarily reflect exactly, the intrinsic properties of the viruses themselves and that such tests are no more than aids in the characterization of the viruses. However, provided that it is accepted that the determinations provide only approximate data for comparative purposes only, they may still be of value (Hill, 1984).

3.1.3.1 Particle Measurements

Visual observation of the shape and size of the virus particle is a basic requirement for identification. In many instances, it may provide a rapid method for identifying the group to which an unknown virus belongs. In the case of rod shaped and filamentous virus particles, length and morphology are particularly characteristic of specific taxonomic groups (Brandes and Bercks, 1965; Walkey, 1985). Methods used to measure particle lengths normally use the negative stain procedure (Noordam, 1973). The use of Immuno-Sorbent Electron Microscopy (I.S.E.M) to differentially trap virus particles is quite useful in situations where virus concentration is low since it markedly increases the number of particles observed under the electron microscope. Arunasalam (1987) found that it increases the detection of CyMV and ORSV, a hundred fold. It does not seem to affect particle integrity and therefore is an ideal method to use when particle measurements are made from plant sap. The magnification at which viruses are photographed are usually calibrated with diffraction grating replicas (Hamilton *et al*, 1981), crystals such as catalase (Wrigley, 1968) or virus particles of known length such as Tobacco Mosaic Virus Bos, 1975).

3.1.3.2 Thermal Inactivation Point

The thermal inactivation point of a virus is generally regarded as the temperatures between which the virus is completely inactivated following an exposure of 10 minutes in crude infectious plant sap (Hill, 1984). The TIP has been used in the early years for virus identification, but the test is affected by factors that influence infectivity (3.1.1) and the result is usually quoted as the limits between which infectivity is lost. It has been suggested that use of the 'half-life' of infectivity will give greater precision to this method. However, in practice it may not be worth adopting this method, which also has its limitations, and recognize that the results for different viruses are only roughly comparable (Gibbs and Harrison, 1976). With some viruses, such as Tobacco Mosaic Virus and Potato Virus X, heat inactivation appears to be closely associated with denaturing of the virus protein (Matthews, 1981). TIP for Potyviruses is generally found between 50°C and 75°C, but is usually found between 55°C and 60°C (Hollings and Brunt, 1981a). Knowledge of the TIP can be quite useful in tissue culture programmes aimed at eliminating the virus from plant tissue using thermotherapy to produce virus free plantlets.

3.1.3.3 Longevity *in vitro*

The Longevity *in vitro* refers to the time that the virus, in crude plant sap, will remain infectious at room temperature (Gibbs and Harrison, 1976; Hill, 1984). It is an approximate test because room temperature varies (Noordam, 1973). Different viruses behave differently when allowed to age *in vitro* at room temperature. Loss of infectivity, in this case, may result from the action of microorganisms or oxidative reactions. Small differences in longevity should not be considered significant. In general, longevity tests should be considered as doing no more than characterizing a virus in terms of whether its longevity *in vitro* can be measured in minutes, days, weeks, months or years (Ross, 1964). Longevity in sap for potyviruses is generally from 1-50 days, but usually 2-4 days (Hollings and Brunt, 1981a). This characteristic is quite useful to know when purification of viruses are aimed at acquiring infective particles, especially for viruses that remain infective at room temperature for a few hours only.

3.1.3.4 Dilution End Point

Dilution end point refers to the point at which sap from infected plants fail to infect mechanically inoculated test plants when diluted increasingly, and is often quoted as the limits between which infectivity is lost (Gibbs and Harrison, 1976). Dilutions are usually on a logarithmic scale but other dilutions have also been used. Like the TIP and LIV tests on properties of viruses in crude sap, this test also is rather crude and is subject to several sources of variation (Ross, 1964) as discussed in section 3.1.1. The dilution end point for potyviruses generally lie between 1/100 and 1/1,000,000 but usually between 1/10,000 and 1/ 100,000 (Hollings and Brunt, 1981a).

3.2 MATERIALS AND METHODS

3.2.1 Host Range

Plants to be used for inoculation were put into a dark box for 24 hours prior to inoculation to increase susceptibility (Matthews, 1981). Carborundum powder (600 mesh) was lightly sprinkled on plant leaves. VPV infected Vanilla and *Nicotiana benthamiana* leaves were ground in 0.01 M Sorensen's phosphate buffer ph 7 or ph 8 using a pestle and mortar. Plant leaves were inoculated with 1:1 and 1:10 dilutions. All inoculations were done at 6 pm in the afternoon to minimize variation due to time of day (Matthews, 1981). Controls were inoculated with buffer only. Plants were examined at two day intervals for 4 weeks and the symptoms recorded.

3.2.2 Physical Properties

3.2.2.1 Particle Measurements

3.2.2.1.1 Grid Preparation

A crystallising dish (90 mm diameter) was filled with water until a convex was formed at the brim. A piece of clean paper (3cm x 25cm) was held at both ends and gently wiped over the surface 3-4 times to remove any dust or dirt particles on the water surface. A drop of 1% parlodion in amyl acetate, was placed in the center of the water. The dish was completely covered with a large beaker for 5 minutes to avoid draughts from rippling the parlodion film forming at the water surface. The beaker was tilted once or twice to allow the evaporating amyl acetate to escape. When the surface of the parlodion exhibited a silvery color, when observed from a 45° angle, the film was ready and the beaker removed. Carbon grids (400 mesh) were placed on the film, dull side down, with a pair of fine forceps. The grids were removed by placing a piece of parafilm, clean side down, on top of the grids and lifted gently. Excess water was then drained off with torn pieces of filter paper and the grids left to dry under a petri dish. The petri dish was propped up on one side to allow air movement to dry the grids. The grids were carbon coated using a Dynavac CS300 vacuum evaporator.

3.2.2.1.2 Particle length measurements

Three parlodion coated carbon stabilised grids were floated film side down on a 50μl drop of 1:1000 dilution of VPV antiserum for 20 minutes. The grids were washed with 30 drops of 1/2 PBS and floated film side down on 50μl of VPV infected sap (extracted by pestle and mortar and squeezed through two layers of muslin cloth) for two hours. The grids were washed again with 30 drops of 1/2 PBS and stained with 2% phosphotungstic acid (PTA) by putting a drop of PTA on the grid and draining the excess stain off by lightly touching the pointed end of a torn piece of filter paper to the edge of the grid. The grids were left for a few minutes to dry before observing the virus particles at 80 kV in a Phillips 301s electron microscope (E.M.).

A photo of a catalase enzyme crystal was taken at 10,000x magnification. Photos of the virus particles were taken at the same magnification. The distance occupied by 100 lattice lines of the crystal (=870 nm) was measured and fed into a computer memory as the standard and were measured using an Apple Graphics Tablet.

3.2.2.2 Thermal Inactivation Point

Chenopodium amaranticolor plants used in the experiment were left in a dark box for 24 hours prior to inoculation. Plants used were at the 13 leaf stage. The first four true leaves were removed because they were too small. Leaf number 5 from the bottom was counted as leaf number 1. Eight *C. amaranticolor* plants were used. Plants were inoculated according to the Scheme in Figure 3.1.

Figure 3.1 Scheme for Estimating Thermal Inactivation Point										
Treatments		L	Plant Number							
A	Control	e a f	1	2	3	4	5	6	7	8
B	40°C	N	A	B	C	D	E	F	G	H
C	45°C	u	B	C	D	E	F	G	H	A
D	50°C	m b	C	D	E	F	G	H	A	B
E	55°C	e	D	E	F	G	H	A	B	C
F	60°C	r	E	F	G	H	A	B	C	D
G	65°C		F	G	H	A	B	C	D	E
H	70°C		G	H	A	B	C	D	E	F
			H	A	B	C	D	E	F	G

VPV infected *N. benthamiana* leaves were ground 1:10 (w/v) in 0.01 M Sorensen's phosphate buffer pH 8, using a pestle and mortar and the sap squeezed through 4 layers of muslin cloth. Two mls of sap was pipetted into each of 2.5 ml freeze drying ampoules (100 x 6 mm), labelled from 40°C to 70°C at 5°C intervals and submerged in a waterbath at the relevant temperature for 10 minutes. The ampoule was quickly cooled in running tap water and stored at room temperature ready for inoculation.

Carborundum powder (600 mesh) was sprinkled lightly on *C. amaranticolor* leaves to be inoculated. Each leaf was inoculated by dipping the right fore finger in the prepared VPV infected *N. benthamiana* sap, supporting the leaf with the palm of the left hand and stroking the leaf surface with the right forefinger twice from the petiole to the tip. Uninfected sap was used as the control. The inoculated leaves were washed with distilled water after 5 minutes and transferred to a cool place away from direct sunlight. The number of lesions produced, on each treated leaf, was recorded after 12 days.

3.2.2.3 Longevity in vitro (LIV)

Two mls of sap prepared as described in 3.2.2.2 was pipetted into each of 2.5 mls freeze drying ampoules labelled as time 0, 2, 4, 6, 8,10, 12, 22, 26,30, 34, 48,72 and 96 hours. Eight leaves on each of 2 C. amaranticolor plants were inoculated according to the scheme in Figure 3.2 using the method as in 3.2.2.2 with the appropriate treatment at the relevant time. The number of lesions, produced after 12 days on each leaf, was then recorded.

Figure 3.2 Scheme for estimating longevity *in vitro*													
Hours	0	2	4	6	8	10	12	22	26	30	34	48	72	96
Number of Plants	2	2	2	2	2	2	2	2	2	2	2	2	2	2

3.2.2.4 Dilution End Point (DEP)

Five grams of *N. benthamiana* leaves were ground in 20 mls, 0.01 M Sorensen's phosphate buffer, pH 8, and appropriate dilutions were made. Plants were inoculated according to the scheme in Figure 3.3 and the method in 3.2.2.2. The number of lesions produced on each inoculated leaf was recorded after 12 days.

Figure 3.3 Scheme for estimating dilution end point (after Roberts, 1964)								
Dilutions				Plant Number				
				1	2	3	4	5	6
A	1:5	L	1	A	D	B	E	C	F
B	1:50	e a	2	E	B	D	A	F	C
C	1:500	f N	3	D	A	F	C	E	B
D	1:5000	u m	4	C	E	A	F	D	D
E	1:50000	b	5	B	F	C	D	A	E
F	buffer	e r	6	F	C	E	B	D	A

3.3 RESULTS

3.3.1 Host Range

Nicotiana benthamiana developed a faint systemic veinal chlorosis after inoculation with VPV infected Vanilla sap in Sorensen's phoshate buffer 0.01 M, pH 7 and 8 (Plate 6). *Chenopodium amaranticolor* developed chlorotic, pinpoint local lesions when inoculated with inoculum in pH 8 buffer. Other test plants used neither developed symptoms nor tested positive to VPV using ELISA (see Table 3.1).

Plate 6. Faint veinal chlorosis on *Nicotiana benthamiana* leaf (A). Compare with non-symptomatic leaf (B).

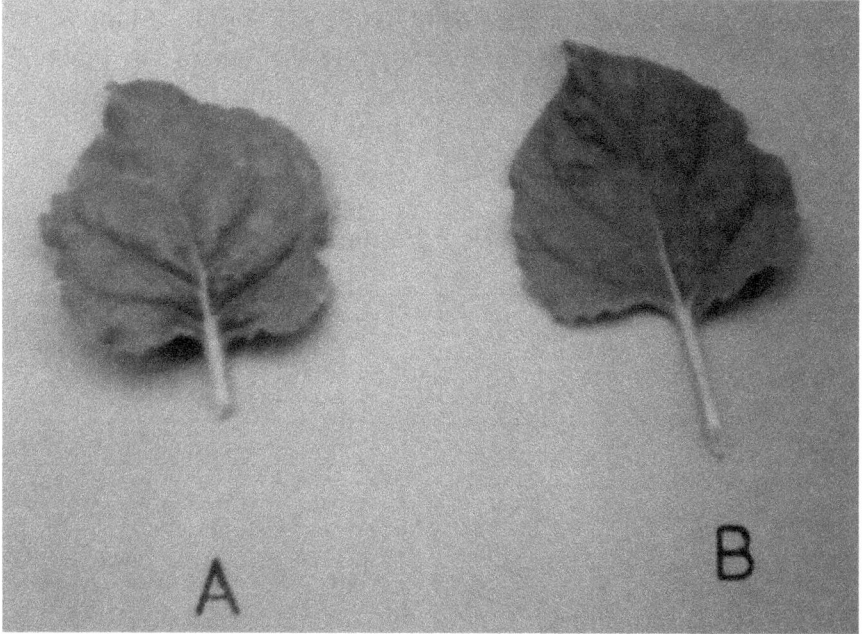

Note: *Nicotiana benthamiana* was the only indicator plant that can be infected directly from VPV infected Vanilla sap, in this study, which produced a systemic infection. *C. amaranticolor* produced pinpoint lesions.

N. benthamiana was an excellent VPV source plant for purification purposes according to Dr M.N.Pearson and Dr A.A. Brunt's study (1990).

Table 3.1 Plants mechanically inoculated with VPV infected Vanilla leaves in 0.01 m Sorensen's phosphate buffer, pH 7 and Ph 8.

Plant (pH 7)	Symptom
1. *Chenopodium amaranticolor*	none
2. *C. quinoa*	none
3. *C. murale*	none
4. *C. occidentalis*	none
5. *C. tora*	none
6. *Datura stramonium*	none
7. *Gomphrena globosa*	none
8. *Nicotiana tabacum cv Xanthi-nc*	none
9. *Nicotiana tabacum cv White Burley*	none
10. *N. glutinosa*	none
11. *N. Clevelandii*	none
12. *N. benthamiana*	Faint systemic veinal chlorosis after 8-10 days
13. *N. megalosiphon*	none
Plant (pH 8)	**Symptom**
1. *C. amaranticolor*	Local chlorotic pinpoint lesions (1mm) diameter after 6-10 days
2. *N. benthamiana*	Faint systemic veinal chlorosis

Plate 7. Faint diffuse mottling on _Nicotiana clevelandii_ leaf (B). Compare with non-syptomatic leaf (A).

N. clevelandii developed a faint systemic mottle when inoculated with VPV infected _N. benthamiana_ sap in Sorensen's phosphate buffer at pH 7 and 8 (Plate 7).

C. amaranticolor developed local, chlorotic, pinpoint lesions, when inoculated with the same inocula (Plate 8). _C. quinoa_ also developed larger local chlorotic lesions (2 mm diameter) when inoculated with VPV infected _N. bethamiana_ sap in 0.01 M Sorensen's phosphate buffer pH 7 and pH 8 (Plate 9). Other test plants used neither developed symptoms nor tested positive to VPV (Table 3.2).

Plate 8. Chlorotic, pinpoint local lesions on *C. amaranticolor* leaf.

Plate 9. Chlorotic local lesions on *C. quinoa* leaf.

Table 3.2 Plants mechanically inoculated with VPV infected *N. benthamiana* leaves in 0.01 M Sorensen's phoshate buffer, pH 8.	
Plant	Symptom
1. *C. amaranticolor*	Local chlorotic lesions (1 diameter) after 6-10 days.
2. *C. quinoa*	Local chlorotic lesions (2 diameter) after 6-10 days.
3. *N. clevelandii*	Faint systemic mottle after 10-12 days.

3.3.2 Physical Properties

3.3.2.1 Particle Measurements

Virus particles observed (Plate 10) were flexuous, filamentous particles with a mean particle length of 776 nm and standard deviation of 31 nm. The particle length ranged from 684 to 870 nm with a median length of 775 nm (Figure 3.4). One hundred per cent of the particles lay between 680 nm and 870 nm which are within the potyvirus length of 680-900 nm as ruled by the International Committee for the Taxonomy of Viruses (I.C.T.V.) (Matthews, 1982).

A Potyvirus Group ELISA Test was available from Sigma USA which works for ALL known potyviruses was used in my work during 1989-1991. We tested weeds and crop plants, on Tongatapu Island, Kingdom of Tonga, for the presence of potyviruses. I don't have those records anymore, but from memory we detected Bean Common Mosaic Virus in some leguminous weeds. Some of the Vanilla plants that tested negative to the VNPV antiserum tested positive to the Potyvirus Group Test. The Potyvirus Group ELISA test was also used for testing large numbers of ZYMV infected squash pumpkin and other cucurbits. A small percentage of these plants were also affected by Watermelon Mosaic Virus II. A paper of this work was published by a magazine for Pacific Island Plant Protection in Australia, in the early 1990s, but it (the magazine) failed to get financial support and was discontinued.

Plate 10. The potyvirus (VPV) found in *V. fragrans*.

This photo was from ISEM using Vanilla sap as the source of Potyvirus particles. Note the differences in appearance of the potyvirus with the photos below probably due to the different preparations for microscopy. The bar is 200 nanometres.

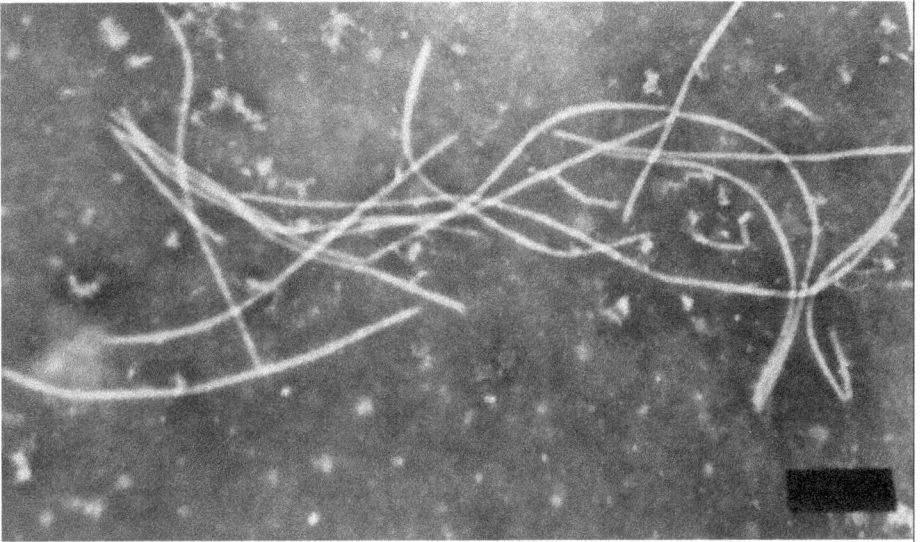

The photo in this box was taken using ISEM with partially purified Vanilla Potyvirus preparations intended for antibody production in a rabbit. Note the broken pieces of the virus particles and the negative stain. Photo was taken by the author on an Electron Microscope at Auckland University, 1988.

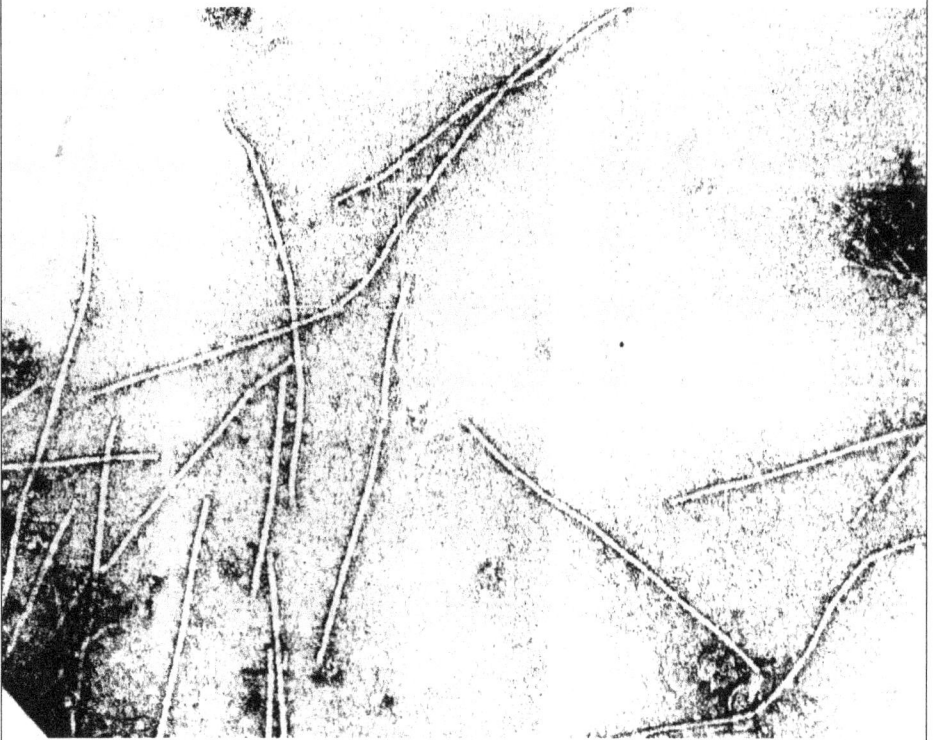

The photo in this box is of the potyvirus   provided by Dr M.N. Pearson. It was taken from purified preparations using *Nicotiana benthamiana* infected plants as the source of virus. This study is reported in Pearson *et al* (1990).

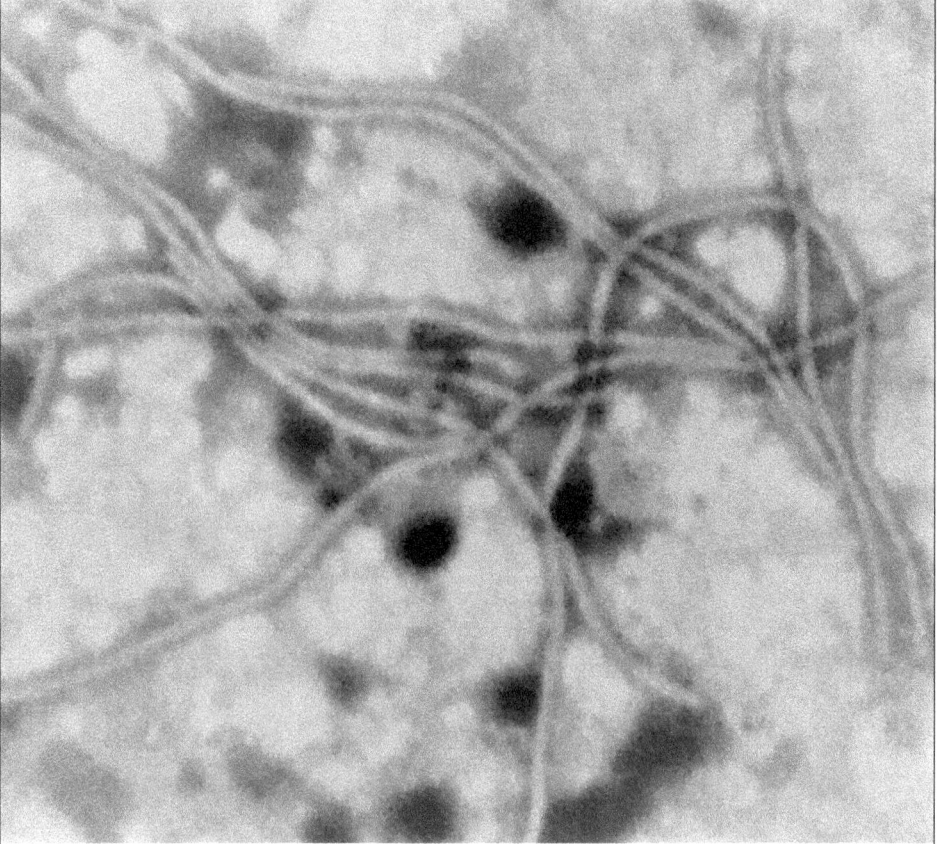

Plate 11. Pinwheel cytoplasmic inclusions induced by VPV in *N. benthamiana* leaves (cross section). Bar = 200 nm. (Photo courtesy of Dr M.N.Pearson).

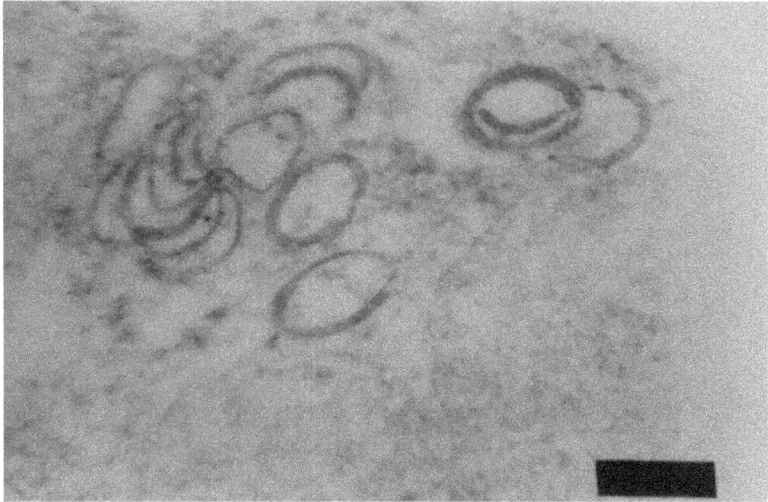

Figure 3.4 VPV particle measurement from Vanilla sap

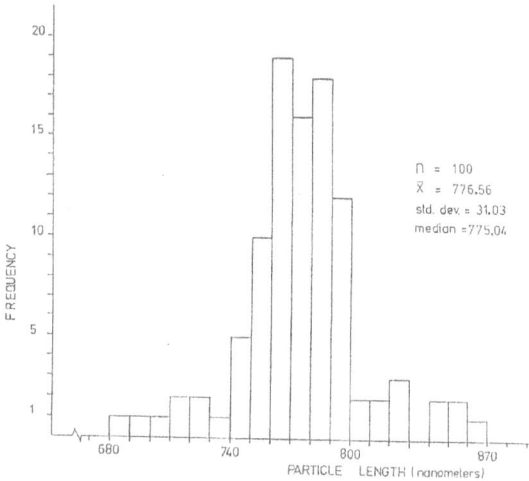

3.3.2.2 Thermal Inactivation Point.

The infectivity of VPV decreased with increasing temperature above 44°C with the TIP between 58°C and 63°C (see Figure 3.5).

Figure 3.5 Effect of temperature on VPV infectivity, in *N.benthamiana* sap, as indicated by local lesions produced in *C.amaranticolor* leaves.

Temperature (°C)	Mean lesion number	standard deviation
untreated	8.25	5.57
39	8.625	5.75
44	8.25	6.13
49	6.25	4.23
54	4.5	5.8
58	0.5	0.75
63	0	0
69	0	0

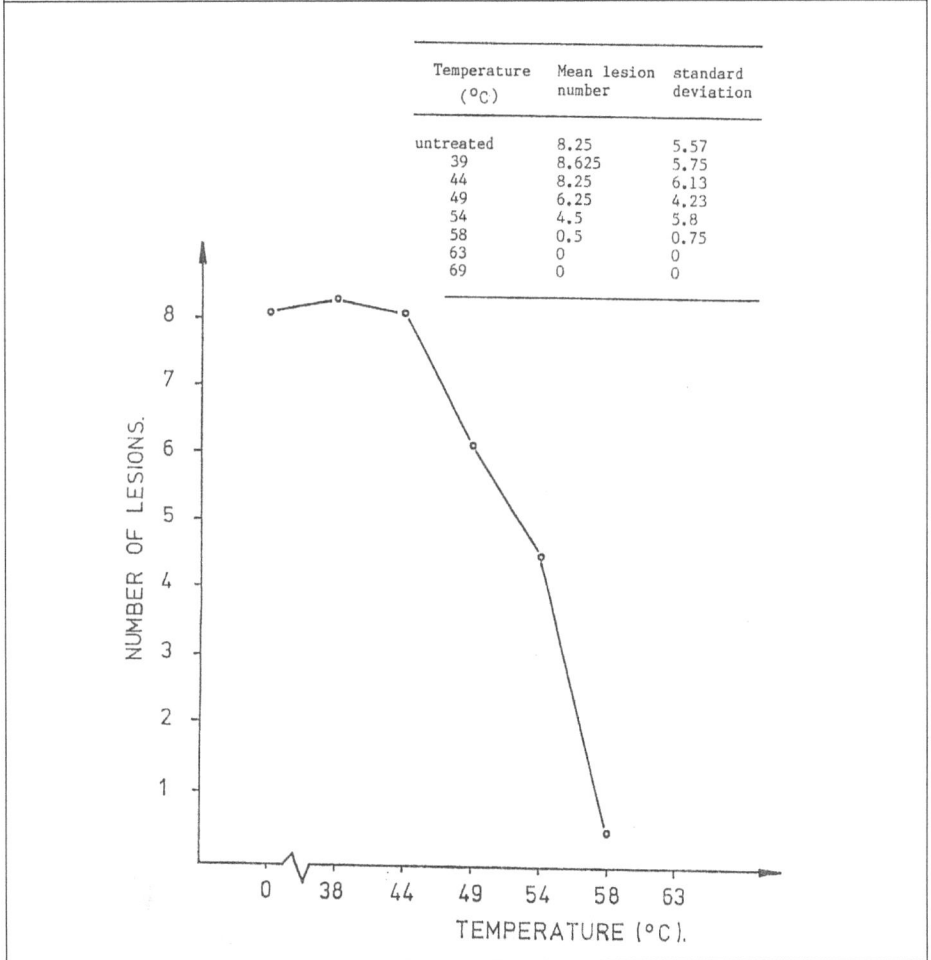

3.3.2.3 Longevity in vitro.

The number of lesions produced decreased with time. The number of lesions produced after 30 hours was less than one per leaf.   It appears that VPV loses most of its infectivity in the first 2 hours (See Figure 3.6).

Figure 3.6 Effect of aging on VPV in *N.benthamiana* sap, as indicated by the number of local lesions produced in *C. amaranticolor* leaves.

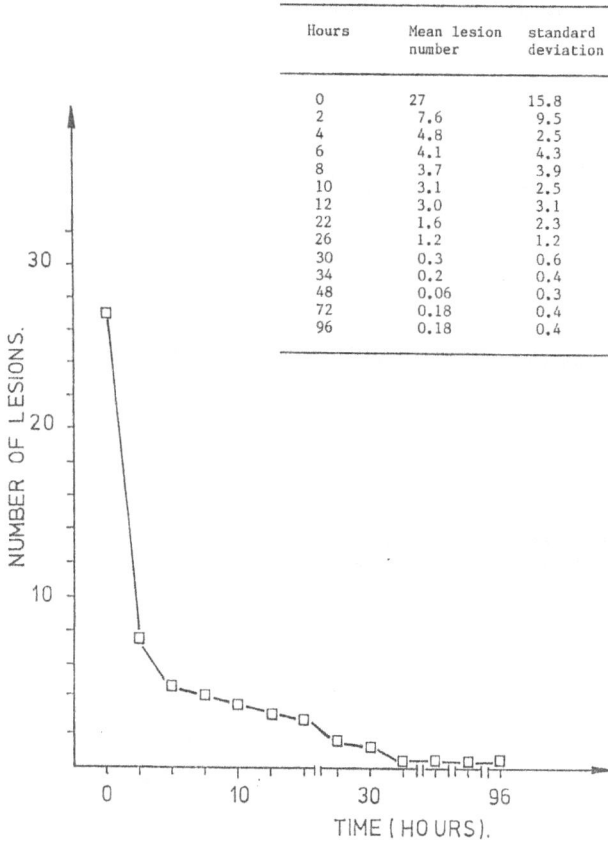

Hours	Mean lesion number	standard deviation
0	27	15.8
2	7.6	9.5
4	4.8	2.5
6	4.1	4.3
8	3.7	3.9
10	3.1	2.5
12	3.0	3.1
22	1.6	2.3
26	1.2	1.2
30	0.3	0.6
34	0.2	0.4
48	0.06	0.3
72	0.18	0.4
96	0.18	0.4

3.3.2.4 Dilution End Point.

The number of lesions produced decreased with the higher dilutions and infectivity appear to be lost at about 1/50,000 dilution (see Figure 3.7).

Figure 3.7 Effect of diluting VPV infected *N. benthamiana* sap on the number of local lesions produced in *C. amaranticolor* leaves.

Dilutions	Mean number of lesions	standard deviation
A – 1/5	178.3	131.6
B – 1/50	95.2	50.7
C – 1/500	8.3	6.6
D – 1/5,000	2.2	2.5
E – 1/50,000	0.5	0.8
F – buffer	0	0

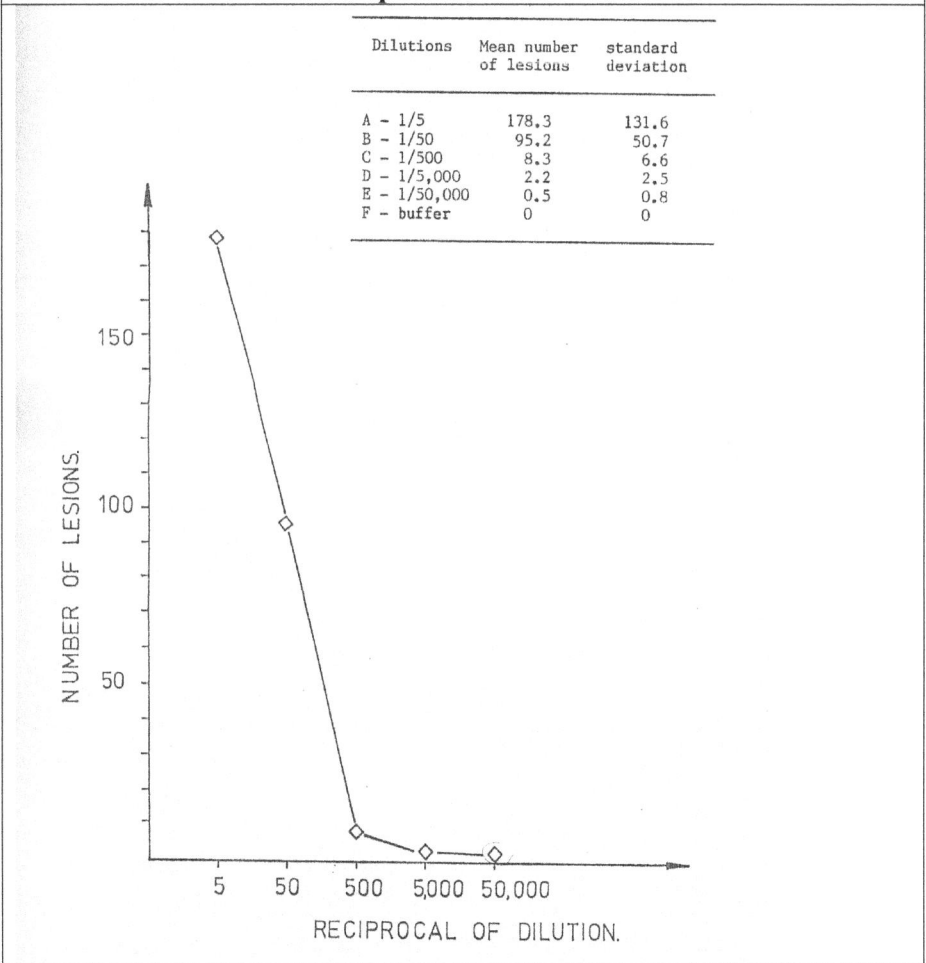

3.4 DISCUSSION

3.4.1 Host Range

Vanilla sap may have contained an inhibitor which prevented the infection of *N. clevelandii*, and the formation of local lesions on *C. amaranticolor* and *C. quinoa* at pH 7. The production of lesions in *C. amaranticolor* using a high pH (8), low molarity (0.01 M) appear to support this conclusion (Table 3.1 and Table 3.2). Hill (1984) states that the inhibition activity of tannins in the plant sap can usually be overcome with a high pH buffer (pH 8-9). This may indicate that tannins are present in the Vanilla sap and prevent lesion formation when lower pH buffers are used. However, since other factors such as temperature, time of day, virus concentration in the sap, other inhibitors eg. ribonucleases, water supply, position of the leaf, nutrition of the plant, light intensity and genetic variation in the host all influence the number of lesions produced (Matthews, 1981), it is quite difficult to attribute the inhibition of lesion formation to any one cause. Additives to the inoculation buffer are sometimes used to overcome the effect of the inhibitors. For example, the addition of sodium sulphite to prevent the action of polyphenoloxidases. Dilution of the sap is also used to overcome the inhibitor effects (Hill, 1984). Another possible explanation is the low concentration of the VPV in the Vanilla sap since the number of lesions produced in *C. amaranticolor* were quite low (<10 lesions, n = 8).

The susceptibility of *N. benthamiana* to VPV in Vanilla sap suggest there may be a chemical compound in this plant to counteract the effect of the inhibitor in Vanilla sap. This chemical compound may not be present in *N. clevelandii* and thus infection was inhibited in this plant. The success of the inoculations from *N. benthamiana* to *N.clevelandii*, *C. amaranticolor* and *C. quinoa* suggest there is no inhibitor present in this plant and is also evidence of the inhibition effect of the Vanilla sap. The production of local lesions on *C. amaranticolor* when a higher pH was used support this conclusion. Undoubtedly, these areas need some more investigations. Further inoculations of these plants and other untested potential hosts, from Vanilla, should be investigated using a high pH, low molarity buffer.

3.4.2 Physical Properties

3.4.2.1 Particle Length Measurements

Particle lengths quoted for a filamentous virus is usually the most common length observer for that virus. Depending on which method is used to isolate the virus, particle lengths may sometimes vary (Walkey, 1985). This variation in particle length is shown by the difference in the particle length measurements using ISEM and those from purified preparations (Section 4.3.1.2). Purification of potyviruses usually result in variable particle lengths. For example, Horner (1986) found that different purification methods result in different particle lengths of a potyvirus infecting *Polianthes tuberosa* L.. Differences in particle lengths for ORSV have been reported by various workers presented in Edwardson and Zettler (1986). However, the particle measurements shown in Figure 3.4 fall within the accepted limits for potyviruses (Matthews, 1982). This suggest that particle integrity is not affected by I.S.E.M.

3.4.2.2 Thermal Inactivation Point

The number of lesions produced to the untreated control was quite low. This is perhaps due to the length of time it took to treat all the ampoules which was about 2 hours. From the Longevity *in vitro* study, we know that 70% of the infectivity of VPV is lost in the first 2 hours (Figure 3.6), therefore the two hours spent in treating all the ampoules, in this experiment, may have contributed to the rapid loss of infectivity and hence the low number of lesions produced. However, the result show that infectivity was lost between 58°C and 63°C (Figure 3.5). The loss of infectivity within these temperatures is consistent with that of potyviruses.

3.4.2.3 Longevity *in vitro*

A 70% reduction in the number of lesions occurred in the first two hours, at room temperature (23°C - 25°C), and slow after that. The effect of lowered temperature and stabilising additives may increase the longevity of the virus, but their effect have to be investigated first. However, possible effects of extraction buffers, clarification agents and other chemicals used in purification may cause the virus to lose infectivity as well. The mean number of lesion per leaf, after 30 hours was less than one. It appears that the infectivity of the virus is largely lost within two days, but complete inactivation may take 4-5 days (Figure 3.6). This result is consistent with loss of infectivity *in vitro* for potyviruses.

3.4.2.4 Dilution End Point

The dilution end point appear to be 1/50,000 which is within the accepted limits for potyviruses (Figure 3.7). The inoculum used for this test was from a different *N. benthamiana* plant than that used for the thermal inactivation and longevity *in vitro* tests. This plant seem to have a higher concentration of virus. Differences in VPV concentration, between the two source plants demonstrate one of the sources of variability in the result from such tests.

3.5 CONCLUSION

The morphology of VPV places it in the potyvirus group. This conclusion is further supported by restricted host range which is typical of potyviruses. VPV can be mechanically transmitted from *Vanilla fragrans* to *N. benthamiana* and *C. amaranticolor*. From N. benthamiana it was transmitted to *C. amaranticolor*, *C. quinoa* and *N. clevelandii* (Section 3.3.1).

VPV was also transmitted by *Myzus persicae* Sulz. and *Aphis gosypii* Glover, in a non-persistent manner, from infected to healthy *N. clevelandii* plants. In addition, VPV induced pinwheel inclusions, typical of potyviruses, were found in VPV infected *N. benthamiana* leaf sections (Plate 11). Furthermore, SDS denatured particles of VPV produced a single major polypeptide band of molecular weight 32,700 which is consistent with that of potyviruses (Pearson, Pone and Brunt, 1990).

As discussed in Section 3.1.1, these characteristics supports our conclusion that VPV is a potyvirus.

The antiserum (Chapter 4) to VPV has been tested against 23 other common potyviruses of monocots from the reference list of the Institute of Horticultural Research, Littlehampton, United Kingdom using I.S.E.M. Results for these tests were negative. Therefore, there is no indication of any serological relationships with other common potyviruses of monocots (Pearson, Pone and Brunt, 1990.)

The evidence therefore suggest that the virus referred to as VPV is definitely a potyvirus and may be an uncited one.

....CHAPTER 4....

DEVELOPING A DETECTION METHOD FOR VPV, CyMV AND ORSV.

4.1 INTRODUCTION

4.1.1 General comments on detection methods

Correct identification of a virus causing disease in the field is essential for the development of adequate control methods (Walkey, 1985). Furthermore, the ability to test for viruses is an essential requirement for their isolation and study. Four general methods for virus assay are available; (i) Infectivity, (ii) Serological, (iii) Physical and (iv) Chemical (Matthews, 1981). Serological techniques are frequently favoured because of their specificity, speed and the scope they provide for standardization. For many important viruses, conventional serological techniques cannot be used because of limitations such as low virus concentration, unsuitable particle morphology or the presence of virus inactivators or inhibitors in plant sap extracts. The limitation can largely be overcome by the use of the microplate method of Enzyme Linked Immuno-Sorbent Assay (ELISA) (Clark and Adams, 1977; Clark 1981).

The probability of detecting a particular virus in infected plant material depends largely on the sensitivity of the test used. In a comparison of several tests ((i) Immuno-osmophoresis (ii) Electron microscopy (iii)

Host infectivity (iv) ELISA and (v) ISEM) used for ORSV and CyMV detection, it was found that ELISA and ISEM were the most sensitive. With these methods, it was possible to detect 2-5 ng/ml and 1 ng/ml for CyMV and ORSV respectively (Arunasalam, 1987). Horner (1986) also concluded that ELISA was more sensitive than the microprecipitin test or immuno-osmophoresis in detecting a potyvirus in *Polianthes tuberosa* L. ELISA was able to detect as low as 159 ng/ml compared with about 10 ug/ml for both the microprecipitin test and immuno-osmophoresis. The use of ELISA is quite suitable for large scale routine testing for viruses. Torrance and Jones (1981) reviewed the suitability and sensitivity of several tests ((i) Floccular eg. Microprecipitin test (ii) Gel diffusion eg. SDS Double Diffusion test (iii) Enzyme labelling eg. ELISA) and concluded that DAS-ELISA was the most suitable for large scale indexing of plants because of its sensitivity and reproducibility.

4.1.2 Virus detection using ELISA

In the DAS-ELISA procedure, the wells, in polystyrene microtitre plates, are coated with purified gamma-globulins specific to the virus to be tested. After removing gamma-globulin by washing (3x), the virus is selectively trapped and immobilised by the solid phase specific antibody. Enzyme labelled specific antibody is then reacted with the immobilized virus, and unreacted enzyme-labelled specific antibodies are removed by washing (3x). Retained enzymes are then assayed by adding a suitable substrate. The use of substrates that give colored hydrolysates permits visual quantitative assessment of the presence or absence of the virus in the test sample. As the retention of enzyme is proportional to the amount of immobilised virus, quantitative measures of the immobilised virus is possible by colorimetry (Clark, 1981). The principles of the DAS-ELISA technique are summarised in Figure 4.2.

The development of a virus specific test using ELISA requires the purification of the virus from plant material, development of an antiserum to the specific virus to be tested by injecting of a rabbit with the purified virus, purification of the virus specific antibodies from the rabbit's blood serum, the conjugation of part of the antiserum with an enzyme and optimizing the test performance through finding the best concentration for coating with both gamma-globulins and enzyme conjugate. Therefore an antiserum had to be produced to the potyvirus found in Vanilla. For CyMV and ORSV antiserum was already available. Enzyme conjugates were also produced for the three viruses and ELISA was optimized for each virus.

4.1.3 Purification of VPV

Purification involves the separation of the virus from the plant material. The degree of purity of the virus separation depends on what has to be done with it afterwards. For chemical analysis, a highly purified virus preparation is required, but less pure preparations will often suffice to produce an antiserum (Noordam, 1973). Furthermore, increasingly stringent criteria of purity have shown that even highly purified virus is often contaminated with host derived antigens. Rybicki and von Wechmar (1985) considered it better to immunise the rabbit with a concentrated semi-purified virus preparation and then exhaustively post-absorb the resulting antisera, to remove plant protein antibodies, than to attempt to purify the virus.

There are four general steps in the purification process. First, the tissue is broken up in extraction buffer so the virus can be released from the cells. This is usually done with a blender or pestle and mortar with less stable viruses. The extraction buffer contains additives to help with the stability and release of the virus. Additives include reducing agents such as sodium sulphite and thioglycollic acid to prevent loss of virus through oxidation (Matthews, 1981), urea to prevent particle aggregation (Hollings and Brunt, 1981b), enzymes like cellulose to breakup cells and aid virus release (Johnstone *et al*, 1982) and various other additives depending on purification method, virus or plant.

The second stage is the clarification of the extracted plant material with a clarifying agent. For potyviruses, the most commonly used clarifying agent is chloroform. The chloroform is added to the plant sap in various concentrations ranging from 3% - 50% w/v (Hollings and Brunt, 1981b).

The third stage is the removal of the macromolecular plant material from the aqueous phase by a low speed centrifugation. Some potyviruses can become lost during this stage of the purification either through adsorption to the plant material or poor release from membrane bound bundles (Francki, 1972).

The last stage involves the sedimentation of the virus. Ultracentrifugation is widely used to pellet potyviruses and it is usually at a moderate centrifugal force to minimise disruption and aggregation of virus particles. A centrifugal force of 58,000 g for 90 minutes is considered to be the best. The resulting pellet can be further purified using sucrose density-gradient centrifugation, which get most potyviruses well separated from host constituents (Hollings and Brunt, 1981b).

Procedures for the purification of most known viruses are available, but different individual viruses and sometimes different strains of the same virus may require specifically different treatments at any stage of their purification. Often, the purification of new viruses must be worked out through trial and error (Walkey, 1985).

4.1.4 Serology

Foreign proteins injected into an animal induces the production of antibodies in the lymphatic tissue. Antibodies are proteins that specifically combine with the substance injected and they belong to a group of proteins called gamma globulins. Serum is the fluid that is left after the blood clot is removed. Antiserum is serum that contain antibodies. Antigen is the protein that stimulate the production of antibodies when injected into an animal (Noordam, 1973). The outer coat or capsid of plant viruses consist of protein subunits of a type and in an arrangement peculiar to each virus. The virus particle, therefore, because of its three dimensional shape and its size forms an ideal antigen which on injection into the blood stream of a suitable animal eg. Rabbit, stimulates the production of antibodies. Such antibodies, which are present in the serum fraction of the blood, react only with the stimulating virus or a very closely related virus (Hill, 1984).

The specific component of antisera is made up of globulins and the type and amount of these determine the quality of the antiserum. Three main glasses of immunoglobulins (gamma globulin) known are IgG, IgA and IgM (van Regenmortel, 1982). Generally IgG are the most important in tests commonly used with plant viruses, such as precipitation, agglutination reactions and ELISA. IgM usually appears first, in the blood, but disappears within a few weeks. IgG first appears a few days after IgM and rises to high concentrations, persisting for long periods of time as circulating antibody (Matthews, 1981). IgG is also the simpliest, existing as a Y shaped molecule of which the arm of the Y are the combining sites which are specific for each antiserum. The specificity of the antigen/antibody reaction is the basis of all the serological tests (Hill, 1984).

Antisera to plant viruses are commonly raised in rabbits, both because of their suitability as laboratory animals and also the useful amount of antiserum obtained (Matthews, 1970). Injection of a virus into a rabbit is usually done intramuscularly, subcutaneously or intravenously. A single dose injected intramuscularly is usually more effective than several doses injected intravenously with a comparable quantity of virus. A combination of intramuscular and subcutaneous injections has been used to induce the production of antibodies with an antiserum titre of 1/2048 in 20 days (Noordam, 1973). Test bleeds are made and a suitable amount of blood is taken from the rabbit when the antibody concentration is optimal.

Since the virus preparations used to inject the rabbit contained plant proteins, which are also immunogenic (van Regenmortel, 1966), it was essential that these proteins are absorbed out.

4.1.5 Optimization of DAS-ELISA

Enzyme labelled antibodies have been used for the quantitative determination of specific antibodies in various human and animal diseases (Voller et al, 1976). The principles of this method have been outlined by Engvall and Perlman (1971, 1972). Its use for plant viruses was first described by Voller *et al* (1976) and Clark and Adams (1977). Optimization of ELISA aims at finding the best concentrations to use for coating gamma globulin, enzyme conjugate and the plant sap. Furthermore, the level of detection was also assessed.

4.2 MATERIALS AND METHODS.

4.2.1 Purification of VPV.

Several extraction buffers were trialed to extract and purify the potyvirus from frozen Vanilla leaf tissue using various additives to help with purification. These are summarized in Table 4.1.

4.2.1.1 Particle measurements.

Photos of the partially purified virus particles were taken at 10,000x magnification, in an Electron Microscope, developed to a final magnification of 60,000x. Seventy three particle lengths were measured using a Hewlett Packard 9874 tracing digitizer (Figure 4.6).

Figure 4.6. VPV particle measurements from a partially purified virus preparation (Procedure 10).

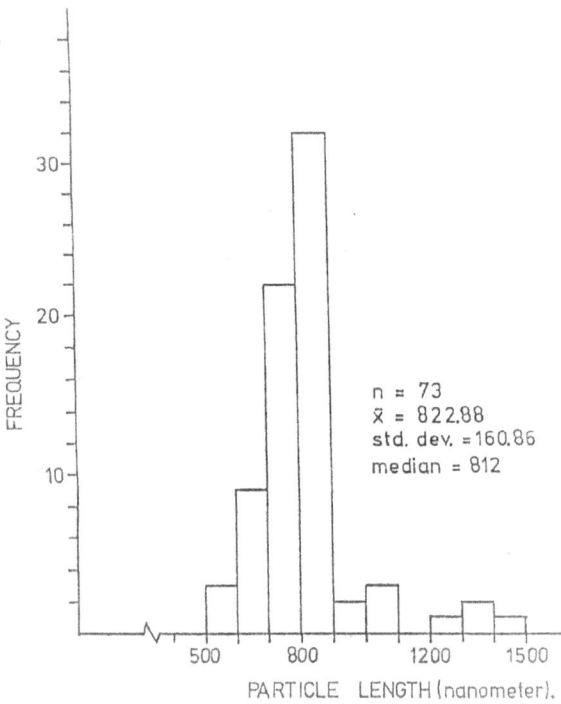

n = 73
x̄ = 822.88
std. dev. = 160.86
median = 812

4.2.1.2 Ultra violet (U.V) absorption.

The absorption spectra of the partially purified virus preparation were scanned from 230 nm to 300 nm and the A260/280 ratios were calculated (Table 4.2).

Table 4.2. A 260/280 ratios of VPV preparation from Procedure 10.			
Normal range for potyviruses	VPV preparation		
1.14-1.25*	A	B	C
	1.17	1.16	1.15
Key : * - Hollings and Brunt (1981a)			

4.2.2 Serology

4.2.2.1 Antiserum production

Partially purified virus, from *Vanilla fragrans* symptomatic leaves, was mixed with an equal amount of Freund's Incomplete Adjuvant (F.I.A.) By repeatedly forcing through a syringe 5-6 times until the creamy emulsion remains intact. A laboratory rabbit was injected with the emulsified preparation. Three injections were made according to the Scheme in Table 4.3.

Table 4.3. Scheme for injection of the rabbit			
Day	Virus concentration	Virus preparation	Injections
1	1.9mg/ml	0.5ml VPV + 0.5 F.I.A.	1/4ml subcutaneously
9	2.24 mg/ml	1ml VPV + 1ml F.I.A.	1/2ml intramuscularly in each leg
16	1.26mg/ml	1ml VPV + 1ml F.I.A.	1/2ml intramuscularly in each leg

Twenty millilitres of blood was bled from the rabbit on day 21 and left to clot for 4 hours at room temperature. The clot was loosened from the wall of the bottle and clear serum was poured off and centrifuged for 30 minutes at 5000g. The pellet was discarded and the serum was stored at 4°C.

4.2.2.2 Testing for Healthy Plant Protein Antibodies (Noordam, 1973).

Since any partially purified virus contain a substantial amount of healthy plant protein, as well as the virus, it is important to test any antiserum for healthy plant protein reaction and absorb them out to minimize healthy background reactions.

Sap extracted from healthy Vanilla tissue and antiserum were diluted in 0.85% NaCl in 0.01 Tris (Tris oxymethylaminomethane) buffer. 10µl of each solution was pipetted into a petri dish according to the scheme in Figure 4.1. Paraffin oil was poured over droplets to prevent evaporation. The petri dish was incubated, in the oven at 37°C for 2 hours. The reactions were observed with a stereo microscope with top light shining down on a dark background. The resulting reactions were evaluated according to the following scale (Table 4.4).

+++ - heavy reaction
++   - moderate reaction
+   - slight reaction
I   - barely visible reaction
-     - no reaction

Figure 4.1 Scheme for the Microprecipitin Tests.								
	Dilutions of healthy or VPV infected sap							
	A	1	1/2	1/4	1/8	1/16	1/32	1/64	1/128
B								
1								
1/2								
1/4								
1/8								
1/16								
1/32								
1/64								
1/128								
1/256								
1/512								
C								

Key - A - healthy sap only (no antiserum added)

B - NaCl only (no healthy or infected sap added)

C - antiserum only (no healthy or infected sap added)

Figure 4.2 Principles of the Double Anti Body Sandwich - Enzyme Linked Immuno-Sorbent Assay (after Clark and Adams, 1977).

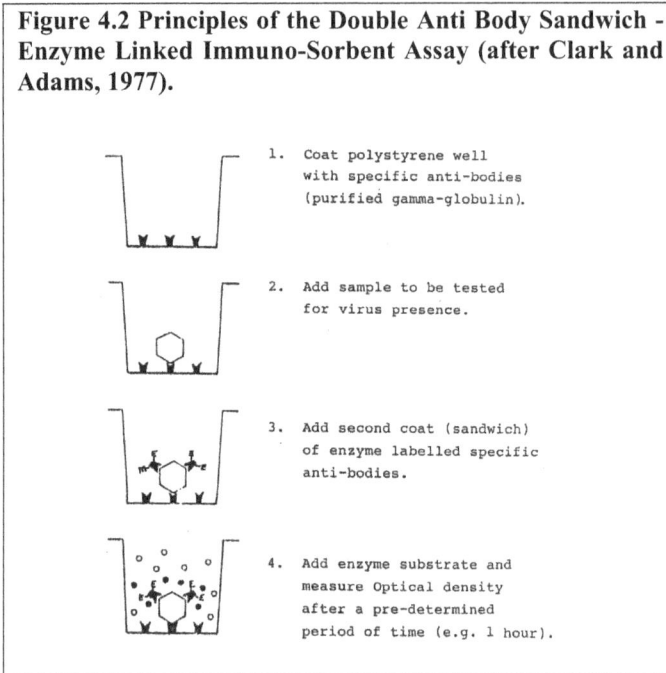

1. Coat polystyrene well with specific anti-bodies (purified gamma-globulin).

2. Add sample to be tested for virus presence.

3. Add second coat (sandwich) of enzyme labelled specific anti-bodies.

4. Add enzyme substrate and measure Optical density after a pre-determined period of time (e.g. 1 hour).

4.2.2.3 Removal of the Healthy Plant Protein Antibodies (Noordam, 1973).

Four volumes of healthy sap [1:5, w/v, in Tris buffer (+ 0.85% NaCl)] were mixed with one volume of antiserum, incubated at 37°C for two hours and transferred to the refrigerator overnight. The antiserum was centrifuged at 5,000g for 30 minutes and the pellet discarded. The supernatant was then frozen overnight, thawed in the morning, and centrifuged again at 5,000g for 30 minutes and the pellet discarded. The pre-absorbed antiserum was tested again for healthy plant protein using the method in 4.2.2.2. The result is presented in Table 4.5.

4.2.2.4 Testing virus infected Vanilla tissue with pre-absorbed antiserum.

Sap extracted from symptomatic Vanilla leaves and antiserum were diluted in 0.85% NaCl in 0.01M Tris buffer. Preabsorbed antiserum was tested for virus antibody reaction using the method in 4.2.2.2. The result is presented in Table 4.6.

4.2.3 Optimization of DAS-ELISA

4.2.3.1 Purification of Gamma Globulins.
(After Clark and Adams, 1977)

Eighteen mls of distilled water was added to 2 mls of preabsorbed antiserum, mixed with 20mls of saturated ammonium sulphate solution and incubated for 1 hour at room temperature. The mixture was centrifuged at 5,000g for 10 minutes and the supernatant discarded. The precipitate was dissolved in 1 ml 1/2 strength PBS and dialysed in 1/2 strength PBS 3x for 2 hrs, 2 hrs and overnight at 4°C. The solution was then passed through a column of pre-equilibrated DEAE cellulose, Sigma No. D-3764. The gamma globulins were eluted from the column with 12 mls 1/2 strength PBS and the effluent was collected in 1 ml aliquots. The Optical Density (O.D.) at 280 nm of each aliquot was measured, using Varian DMS 90 spectrophotometer, and aliquots with the highest absorbance retained. The concentration of these were adjusted to give an A280 reading of approximatedly 1.4 which is equivalent to 1mg/ml gamma globulin concentration. The gamma globulins were stored in silicone coated tubes in 1 ml lots at 4°C, adding 0.01 mg/ml sodium azide for preservation.

4.2.3.2 Conjugation of Gamma Globulins for CyMV, ORSV and VPV.
(After Clark and Adams, 1977)

One ml of alkaline phosphatase (Sigma type VII-S) was centrifuged at 10,000 rpm for 1 minute, the supernatant was discarded and the precipitate dissolved in 2 mls purified gamma globulin. The mixture was dialysed 3 times in 500 mls PBS for 2 hrs, 2 hrs and overnight at 4°C. Fresh electron microscope grade glutaraldehyde solution was mixed with the conjugate to 0.06% final concentration and left for 4 hours at room temperature before dialysing 3 times against PBS to remove excess glutaraldehyde. Bovine serum albumin (Sigma RIA grade) was added to 5 mg/ml final concentration and the conjugate was stored at 4°C.

The test procedure for DAS ELISA, used in this study, is shown in Figure 4.3.

The procedure follows that of Clark and Adams, 1977 (see principles of DAS ELISA in Figure 4.2) but the antibodies used were produced during this study.

Figure 4.3 DAS-ELISA Test Procedure (after Clark and Adams, 1977). All chemicals used for DAS-ELISA are listed in Appendix 1.

Day 1. - Make appropriate dilutions of gamma globulins in coating buffer. One ELISA microtitre plate requires 19.2 mls.
- Add 200µl of coating gamma globulins to appropriate wells in microtitre plates
-Place plate on moist tissue in air tight container for 2-4 hours at 37°C and record incubation time.

Day 1 continued - Preparation of Test Samples.
- Grind leaf samples, with pestle and mortar, in PBS-tween + 2% PVP.
- Leave at 4°C or room temperature for 2 hrs to settle the plant material and make further dilutions as appropriate.

Day 1 continued - Addition of Test samples
-Wash the plate 3x with PBS tween at 3-5 minute intervals.
-Add 200µl of test samples to appropriate wells.
-Place plate in a moist airtight container and leave at 4°C overnight.

Day 2. - Addition of enzyme conjugated gamma globulins.
- Prepare appropriate dilutions of enzyme conjugated gamma globulin in PBS-tween + 2% PVP + 0.2 % ovalbumin.
-Wash plate 3x with PBS-tween.
-Add 200 µl of conjugate dilutions to appropriate wells.
-Incubate plate in a humid air-tight container at 37°C for 2-6 hrs. Record incubation time.

Day 2 continued - Addition of substrate
-Prepare 25 mls of 0.6 mg/ml phosphate substrate per plate (use acid washed glassware).
-Wash plate 3 times with PBS-tween.
-Add 250 µl substrate per well.
- Incubate at room temperature for one hour and observe reaction.
-Read plate on ELISA reader at 405 nm (Biotek Model EL-307)

Scheme for determining the optimum concentration of coating and enzyme labelled gamma-globulin (see Figure 4.4 and Appendix 2).									

Concentration of gamma globulins (µg/ml)									
		10µl			1.0µl			0.1µl		
Test samples	I	1:1								
		1:10								
		1:100								
	II	1:1								
		1:10								
Control Samples		1:1								
		1:10								
PBS Tween										
Enzyme Conjugate		1:400	1:800	1:1600	1:400	1:800	1:1600	1:400	1:800	1:1600

Key - I = high virus concentration test sample
II = low virus concentration test sample

4.2.3.3 Determining of the optimum concentration of coating and enzyme labelled gamma globulin.

(After Clark and Adams, 1977).

The ELISA Test Procedure in Figure 4.3 was used to find the optimal concentrations of the plant sap, purified gamma globulin and enzyme conjugate using the scheme in Figure 4.4. The optimum concentrations are given in Table 4.7. Full results are presented in Appendix 2.

Figure 4.4. Scheme for determining the optimum concentrations of coating and enzyme labelled gamma-globulins									
Concentration of coating gamma globulin (µg/ml)									
		10			1.0			0.1		
Test samples										
Control samples										
PBS-Tween										
	1:400	1:800	1:1600	1:400	1:800	1:1600	1:400	1:800	1:1600	
Dilutions of Enzyme-Labelled conjugate									

4.2.3.4 Determination of assay sensitivity.

To determine the level of virus detectability of the antiserum, a serial dilution of a purified virus preparation was made and tested by ELISA. A comparison with healthy and VPV infected Vanilla leaves was also made in Figure 4.7.

4.3 RESULTS.

4.3.1 Purification of VPV.

4.3.1.1 Procedures for the attempted purification of VPV

The best procedure for purifying VPV from infected frozen Vanilla leaves was procedure 10 (Figure 4.5). Concentrations of virus in the partially purified preparation was estimated from the ultraviolet absorption, at 260 nm, of a 1:10 dilution of the preparation. The results of the other procedures for the purification of VPV are summarized in Table 4.1.

Table 4.1 Purification procedures trialed for VPV from frozen *Vanilla fragrans* leaves from the Kingdom of Tonga.

Extraction buffer and additives	Clarification	Sedimentation	Resuspension buffer	Further clarification	Comments based on Electron Microscope examination
1. 1:2 w/v 0.01 PO$_4$, pH6, 1M Urea, 5%TGA, Cellulase (Sigma Type II)	50% CHCl$_3$ 90 minutes	Ult.centri. 90 mins at 58,000g	0.1MPO$_4$, pH6, 1M Urea 5% TGA	-	No virus particles present
2. 1:2 w/v 0.1MPO4, pH6 1MUrea 5%TGA	50% CHCl$_3$ 90 minutes	Ult.centri. 90 mins at 58,000g	0.1MPO$_4$, pH6, 1M Urea 5% TGA	-	No virus particles present
3. 1:2 w/v 0.1MPO$_4$, pH7 1MUrea 5%TGA	50% CHCl$_3$ 90 minutes	Ult.centri. 90 mins at 58,000g	0.1MPO$_4$, pH6, 1M Urea 5% TGA	-	No virus particles present
4. 1:2, w/v 0.1MPO$_4$, pH8 1MUrea 5%TGA	50% CHCl$_3$ 90 minutes	Ult.centri. 90 mins at 58,000g	0.1MPO$_4$, pH6, 1M Urea 5% TGA	-	No virus particles present
5. 1:2 w/v 0.5MPO$_4$, pH 7.5, 1MUrea 0.04MNa$_2$SO$_3$7 H$_2$O	50% CHCl$_3$ 90 minutes	Ult.centri. 90 mins at 58,000g	0.01MPO$_4$, pH7.5, 0.5MUrea, 0.02M Na$_2$SO$_3$7H$_2$O	-	Many virus particles with lots of plant debris
6. 1:2 w/v 0.5MPO$_4$, pH7.5, 1MUrea 0.04MNa$_2$SO$_3$7 H$_2$O	50% CHCl$_3$ 90 minutes	PEG	0.01MPO$_4$, pH7.5, 0.5MUrea, 0.02M Na$_2$SO$_3$7H$_2$O	-	Many virus particles with lots of plant debris

Table 4.1 Purification procedures for VPV from frozen *Vanilla fragrans* from the Kingdom of Tonga (continued)

7. 1:2 w/v 0.5PO$_4$,pH7.5 1MUrea 0.04MNa$_2$SO$_3$7H$_2$O	50% CHCl$_3$ 90minutes	Ult.centri. 90 mins at 58,000g	0.01MPO$_4$, pH7.5, 0.5MUrea, 0.02M Na$_2$SO$_3$7H$_2$O	10-40% density gradient centrifugation	A few particles
8. 1:2,w/v 0.5MPO$_4$, pH5 1MUrea 0.04MNa2SO 37H2O	80% CHCl$_3$ 120 minutes	Ult.centri. 90 mins at 58,000g	0.01MPO$_4$, pH6.4, 1MUrea, 0.04M Na$_2$SO$_3$7H$_2$O	-	A few particles
9. 1:2, w/v 0.5MPO$_4$. pH6 0.04MNa$_2$SO$_3$7H$_2$O	50% CHCl$_3$ 120 minutes	-	0.01MPO$_4$, pH7.5, 0.5MUrea, 0.02M Na$_2$SO$_3$7H$_2$O 2.5%v/vtween 20	2mls 30% sucrose cushion 58,000g	A few virus particles
10. 1:2, w/v 0.1MPO$_4$, pH6 1MUrea 0.04Na$_2$SO$_3$7H$_2$O	50% CHCl$_3$ 120 minutes	Ult. Centri. 90mins at 58,000g+2mls 25% sucrose cushion	0.01MPO$_4$, pH6, 0.5MUrea, 0.02M Na$_2$SO$_3$7H$_2$O	3mls 25% Sucrose cushion, 58,000g+3000 g for 10 minutes	Many virus particles with less plant debris
11. 1:2, w/v 0.1MPO4 pH6 1M Urea 0.5%TGA 0.04Na2SO37H2O 1%w/v cellulase (Sigma type II)	50% CHCl$_3$ 120 minutes	Ult. Centri. 90mins at 58,000g+2mls 25% sucrose cushion	0.01MPO$_4$, pH7.2, 1MUrea, 0.02M Na$_2$SO$_3$7H$_2$O	3mls 25% Sucrose cushion, 58,000g+3000 g for 10 minutes	Many virus particles with less plant debris

Key: w/v - weight to volume, 1:2w/v - one unit weight (vanilla tissue) to two volumes of extraction buffer, PO4 - Sorensen's Phoshate buffer, PEG - Polyethylene Glycol, TGA - thioglycolic acid, ult. centri. - ultracentrifugation, Na$_2$SO$_3$7H$_2$O - sodium sulphite, CHCl$_3$ - chloroform

Figure 4.5 Final procedure adopted for partial purification of VPV.

1. Weigh out 100 gms frozen VPV infected Vanilla tissue.
2. Cut into thin strips with a sharp razor plate.
3. Pour into a Wareing commercial blender, add 200 mls of extraction buffer (0.1 M Sorensen's phosphate buffer pH 6, 1M Urea and 0.04M sodium sulphite) and homogenise for 3 minutes.
4. Squeeze homogenate through 4 layers of muslin cloth into a large beaker.
5. Measure volume of homogenate and add an equal volume of chloroform.
6. Emulsify in an ice bath for 1-2 hours.
7. Centrifuge emulsion at 3,000 g for 15 minutes.
8. Pipette off and retain supernatant.
9. Centrifuge supernatant at 58,000 g for 90 minutes over a 2 ml, 25% sucrose cushion.
10. Resuspend pellet in resuspension buffer (0.1M Sorensen's phosphate buffer pH 6, 0.5 M Urea and 0.02 M sodium sulphite) overnight at 4°C.
11. Overlay virus suspension from step 10 on 3 mls, 25% sucrose cushion and centrifuge at 58,000 g for 110 minutes.
12. Resuspend pellet in 1x standard saline citrate and centrifuge at 3000 g for 10 minutes.
13. Discard pellet and store virus suspension at 4°C overnight.
14. Centrifuge, again, at 3000 g for 10 minutes.
15. Discard pellet and store virus suspension for injection into rabbit.

4.3.1.2 Particle measurements.

Virus particles observed were flexuous, filamentous particles with a mean particle length of 823 nm and a standard deviation of 161 nm. The particle lengths ranged from 505 nm to 1476 nm with 80% of the particles lying between 680 nm and 900 nm (Figure 4.6).

4.3.1.3 Ultra violet (U.V.) absorption.

The A260/280 of the three preparations (A, B and C) were 1.17, 1.16 and 1.15 respectively (Table 4.2).

Table 4.2 A 260/280 ratios of VPV preparations from procedure 10.			
Normal range for potyviruses	VPV preparations		
	A	B	C
1.14 - 1.25*			
	1.17	1.16	1.15
* Hollings and Brunt (1981a)			

4.3.2 Serology.

4.3.2.1 Testing for healthy protein antibodies.

Reaction to healthy sap was very strong with a titre of greater than 1/512. No reaction was observed in the three controls (A, B and C) (Table 4.4).

Table 4.4 Antiserum reaction to healthy plant sap.							
	Healthy Vanilla Sap						
		A	1	1/4	1/16	1/32	1/64	1/280
Antiserum..Dilution	B		-	-	-	-	-	-
	1/2	-	+++	+++	+++	++	++	+
	1/4	-	+++	+++	+++	++	++	+
	1/16	-	+++	+++	+++	++	++	+
	1/64	-	++	++	++	+	+	I
	1/256	-	++	++	++	+	+	-
	1/512	-	+	I	I	I	-	-
	C		-	-	-	-	-	-

4.3.2.2 Removal of healthy plant protein antibodies.

After removal of the healthy protein antibodies, the reaction to healthy plant sap had a titre of 1/4. No reaction was observed in the controls (A, B and C) (Table 4.5).

Table 4.5 Reaction of healthy sap with preabsorbed antiserum.							
	Healthy Vanilla Sap						
		A	1	1/4	1/16	1/32	1/64	1/128
	B		-	-	-	-	-	-
	1	-	++	++	+	I	-	-
	1/2	-	+	+	+	I	-	-
	1/4	-	I	I	I	I	-	-
Antiserum Dilution	1/8	-	-	-	-	-	-	-
	1/16	-	-	-	-	-	-	-
	1/32	-	-	-	-	-	-	-
	1/64	-	-	-	-	-	-	-
	C	-	-	-	-	-	-	-

4.3.2.3 Testing virus infected tissue with preabsorbed antiserum.

A strong reaction was observed between the antiserum and VPV infected vanilla sap with a titre greater than 1/64. No reaction was observed in the controls (A, B and C) (Table 4.6).

Table 4.6 Reaction of VPV infected Vanilla sap to preabsorbed antiserum.

Antiserum..Dilutions	Infected Vanilla sap dilutions							
		A	1	1/2	1/4	1/8	1/16	1/32	1/64

Antiserum..Dilutions		A	1	1/2	1/4	1/8	1/16	1/32	1/64
	B	-	-	-	-	-	-	-	-
	1	-	++	+	+++	+++	++	++	++
	1/2	-	++	++	++	++	++	++	+
	1/4	-	+++	++	++	+	+	+	I
	1/8	-	+++	++	+	+	+	I	-
	1/16	-	+++	++	+	+	+	-	-
	1/32	-	+++	+	+	I	I	-	-
	1/64	-	++	+	+	I	I	-	-
	C		-	-	-	-	-	-	-

4.3.3 Optimization of ELISA

4.3.3.1 Determination of optimum concentrations of coating and enzyme labelled gamma globulin.

The best concentrations found for the coating gamma globulins and enzyme conjugates were at concentrations of 1µl/ml and 1/1600 respectively (Table 4.7).

Table 4.7 Optical density (405nm) measurements for the optimal concentrations of coating gamma globulin and enzyme conjugates at the VPV infected sap indicated, after 1 hour.

Coating gamma globulin concentration		1 μl/ml	1 μl/ml	1 μl/ml
		VPV	CyMV	ORSV
Test sample	1:1	0.117	0.145	-
	1:10	0.088	0.146	0.198
	1:100	-	-	0.206
Control	1:1	0.013	0.013	0.034
	1:10	0.011	0.013	0.031
PBS tween		0.016	0.011	0.003
Enzyme Conjugate Dilution		1:1600	1:1600	1:1600

4.3.3.2 Determination of assay sensitivity.

The minimum concentration of VPV clearly detectable by ELISA, using the antiserum developed to VPV, is about 85 nanograms (Figure 4.7)

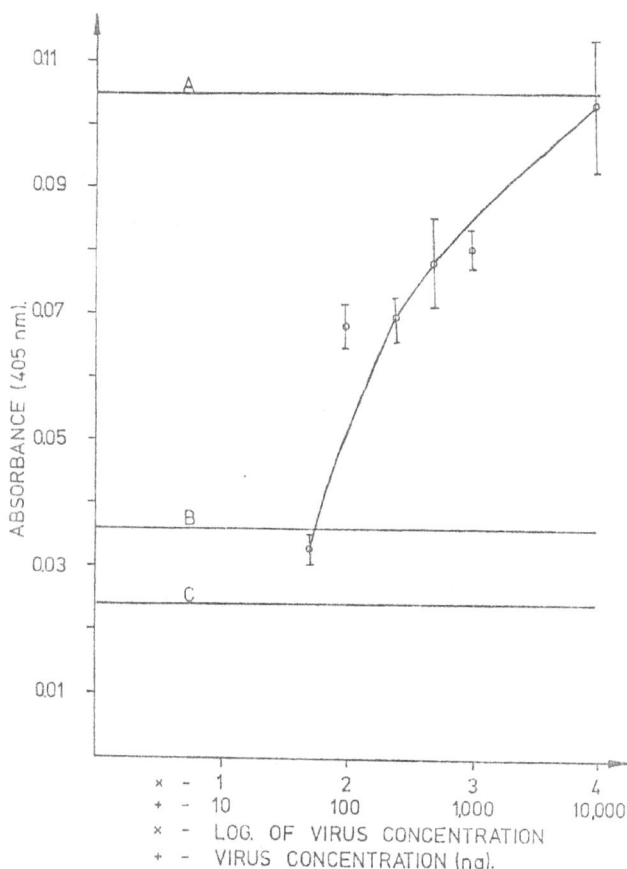

Figure 4.7 Log of VPV concentration (ng) plotted against absorbance at 405 nm after a 1 hour incubation from addition of substrate.
A - VPV infected vanilla sap
B - healthy vanilla sap + 2 standard deviations
C - healthy vanilla sap

4.4 DISCUSSION

4.4.1 Purification of VPV.

All the purification were done from frozen infected, symptomatic Vanilla leaves, except purifications 1-4. The purification of VPV was difficult, which is characteristic of many potyviruses (Hollings and Brunt, 1981b). The greatest problem encountered was the irreversible aggregation of the virus particles. Visual detection of the virus particles from the various preparations, by electron microscopy, found virus particles to be aggregated with bits of plant material. This effect could have been due to the freezing of the leaf tissue rather than an effect of the purification procedure itself. Derrick and Brlansky (1975), quoted by Hollings and Brunt (1981b), found that freezing of tissue or extracts gave massive aggregation of Potato Virus Y, although heating the extracts for 10 minutes at 45°C was found to reverse the aggregation. However, heating the resuspended VPV for 10 minutes at 45°C did not seem to affect the aggregation of this preparation. Hollings and Brunt (1981b) suggested the use of chelating agents such as EDTA (ethylenediaminetetra-acetic acid) for reduction of particle aggregation, but this was not attempted in this study.

The use of the sucrose density gradient (Procedure 7) also did not improve the situation as the aggregated virus particles and the attached plant material sedimented at the bottom of the tube. PEG (polyethylene glycol) precipitation also seem to enhance aggregation. It is known that although PEG-precipitation (4%w/v+4%w/v NaCl) is a comparatively gentle method, it can induce serious potyirus aggregation (Hollings and Brunt, 1981b).

Procedure 10 was the most successful in producing partially purified virus from frozen Vanilla symptomatic leaves. The use of the sucrose cushions, during high speed centrifugation, coupled with low speed centrifugation at 3,000 rpm, appears to be the best method in removing the contaminating plant material. There was still some loss of virus particles during low speed centrifugation since electron microscope examination of grids prepared from the sedimented pellet contained virus particles. Virus yield from this method was estimated in the range of 1.26 mg/ml to 2.9 mg/ml from 100 gms of frozen Vanilla symptomatic leaf tissue, based on the A260 value.

4.4.1.1 Particle measurements

Normally, at least 100 particles of a virus are measured for an estimation of particle length (Walkey, 1985). However, it was not possible to find unaggregated virus particles that could be easily measured. Most unaggregated particles were very short and were not measured since the particle measurements was aimed at estimating the intact particle length.

The particle lengths measured from the partially purified virus (Figure 4.6) preparation appear to be longer than those virus particles trapped from sap extracts by ISEM (Figure 3.1). This could have been due to effects of the purification procedure. The very long particles (>1,000 nm) could have been end to end aggregation of VPV particles. Shorter particles (<700 nm) could have been caused by disruption of the particles during purification. Only unaggregated long particles were measured, therefore this bias in the selection of particles to be measured may have affected the mean particle length obtained.

4.4.1.2 U. V. Absorption

The ultraviolet absorption (220-300 nm) of potyviruses are characteristic for the group (Table 4.2). The potyviruses have a maximum absorption at 260-262 nm and minimum values at 240 - 246 nm and an A260/280 ratio between 1.14-1.25 (Hollings and Brunt, 1981a). The U.V. absorption spectra of VPV did not show a clear maximum or minimum absorption over the range measured although the A260/280 ratios of the partially purified virus preparation have typical potyvirus values. The presence of plant contaminants (Plate 6) probably did affect the absorption values. These values were not corrected for light scattering, therefore the 'true' values after light scattering correction could have been different.

4.4.2 Serology.

4.4.2.1 Antiserum production.

The healthy reaction to the antiserum was initially quite high (Table 4.4). However, after removal of the healthy protein antibodies this reaction was much reduced (Table 4.5). Testing the pre-absorbed antiserum with VPV infected sap showed that it had a titre of greater than 1/64 (Table 4.6). It is quite important that the healthy plant protein reaction be absorbed out of any antiserum to minimize background reaction. The procedure used to absorb out the antibodies to the healthy plant protein in this study appear to work quite well. The reduction in the amount of healthy reaction (Table 4.5) is good evidence of the low level of healthy reaction after absorption. The dilution used in the ELISA procedure (Table 4.7) proved this low level of healthy protein is quite negligible, as there was a clear distinction between the healthy A405 readings and the VPV infected Vanilla sap.

4.4.3 Optimization of ELISA

4.4.3.1 Determination of optimum concentrations of coating and enzyme labelled gamma-globulins.

This procedure is quite straightforward as it is routinely used. The antisera to all three viruses produced clear differences between the virus positive samples (Table 4.7). The lower readings for the low positive samples (Appendix 2) demonstrates the limitations of ELISA in testing very low virus concentrations. However, as discussed in section 4.1.1 ELISA, despite its limitations, appears to be one of the best virus tests available. The lower ELISA readings for VPV compared to CyMV and ORSV could have been due to the difference in the concentration of the three viruses in the infected sap used or low titre of the antiserum.

4.4.4 Determination of assay sensitivity.

The sensitivity of the ELISA test using the antiserum developed to VPV appear to be about 85 nanograms virus per ml (Figure 4.7). The relatively low ELISA readings could have been due to the poor stability of the virus since the purified virus used for this test had been stored at 4°C for 5 days before it was used. As demonstrated by the longevity *in vitro* test, VPV lasts only for a short period in *Nicotiana benthamiana* sap at room temperature, with 70% of its infectivity lost in the first two hours (Figure 3.6). Another possible reason for the low readings could have been low antibody titre. Subsequent ELISA tests carried out with another antiserum developed to VPV by a 'booster' injection, on the same rabbit used to develop the antiserum used in this study, showed much higher readings when tested against VPV infected Vanilla sap (Pearson pers. comm). However, there was a clear difference between the VPV containing samples and the controls using the antiserum tested here (Table 4.7) which proved sufficient for routine testing of Vanilla samples for VPV presence.

....CHAPTER 5....

PATHOGENICITY

5.1 General Introduction

When a microorganism is found on a diseased plant, it is usually identified by reference to special manuals; if the microorganism is known to cause such a disease, then the diagnosis of the disease may be considered complete. If the microorganism found seems to be the cause of the disease, but no previous reports exist to support this, then Koch's postulates are usually applied to prove the pathogenicity of the organism.

Koch's postulates demand that the organism;

(i) must be found to be associated with all the diseased plants examined, (ii) must be isolated and grown in pure culture on nutrient media, and its characteristics described (non-obligate parasites), and its appearance and effects recorded, (iii) inoculum from pure culture must be inoculated on healthy plants of the same species or variety on which the disease appears , and it must produce the same disease on the inoculated plants, (iv) must be re-isolated into pure culture and its characteristics must be exactly those observed in (ii).

Unfortunately, viruses cannot be cultured and when purified some viruses lose their infectivity. Reproduction of the virus symptoms is often achieved by (a) grafting part of a diseased plant onto a healthy plant, (b) rubbing sap from a diseased plant onto a healthy plant (mechanical inoculation), (c) or by allowing certain insects, or other vectors, to feed on the diseased plant then transmit the virus onto a healthy plant (Agrios, 1978).

Grafting is considered the most universally applicable method of virus transmission, requiring only that the virus become systemic in plants and can be joined by grafting (Fulton, 1964). There are many different ways in which a graft may be established, but mostly involves the union of a shoot portion of one plant, referred to as the scion, and the root bearing portion of another, called the stock. If either the scion or the stock is infected the virus will pass on to the healthy portion and virus infection is established. Grafting was widely used in the early days of plant virology for the transmission of   plant viruses that could not be transmitted in any other way. It is now not used so frequently by Virologist, but it is useful in studying some viruses infecting woody species (Walkey, 1985). Although grafting works well with some plants and is the only method of virus transmission for others, it is generally not an efficient method of virus transmission. The vascular tissue of the stock and scion may not be completely connected (Matthews, 1981) or as in the monocots, grafting is impossible for most species (Fulton, 1964). Sometimes a dormancy period of several months occurs before symptom appearance in many woody plants (Walkey, 1985).

Mechanical inoculation is considered a more efficient method of virus transmission (Matthews, 1981) at least for most viruses. It aims to create a multitude of small wounds on the plant surface without killing the cells so the virus can enter the cells. This is achieved by using abrasives such as carborundum powder by rubbing together with infected sap on the surface of the inoculated leaf.

Plant viruses cause both internal and external symptoms in plants (Smith 1974, Bos 1978). The internal symptoms include the formation of cytoplasmic inclusions (Chapter 3), reduction in chloroplast activity, changes in hormone levels, reduction in availability of fixed carbon and other changes. The more observable effects include stunting, growth abnormalities, hyperplasia (increased growth and differentiation) and hypoplasia (reduced growth and differentiation) (Matthews, 1981). The most common and sometimes the only kind of symptom is reduced growth rate of the plant, resulting in various degrees of dwarfing or stunting of the entire plant. Almost all viral diseases seem to cause some degree of reduction in total yield and the length of life of the virus infected plant is usually shortened. The most obvious symptoms of virus infected plants are usually those appearing on the foliage but some viruses can cause striking symptoms on the stem and the fruits with or without symptom development on the leaves (Agrios, 1978).

Symptoms caused by potyviruses are generally similar in monocotyledonous plants. Infected species usually grow longitudinal chlorotic streaks that may become necrotic (Hollings and Brunt 1981b).

This chapter will deal with the relative effects of CyMV, ORSV and VPV on Vanilla plants. The investigation first tested and compared the relative incidence of each virus in the different symptoms observed on the Vanilla foliage (Tables 5.1, 5.2, 5.3, 5.4). Preliminary conclusions were made concerning the effects of each virus on Vanilla plants which were used as a guideline in the pathogenicity study. To test the pathogenicity of the three viruses on Vanilla plants, healthy Vanilla plants were grafted and mechanically inoculated with infected plant material to confirm the preliminary conclusions (Tables 5.5, 5.6). Aphids were also used to attempt transmissions of VPV from VPV infected vanilla plants to healthy plants. The effect of VPV on the growth rate of infected and healthy plants was also investigated (Figure 5.2).

5.2 MATERIALS AND METHODS

5.2.1 Symptoms and virus presence

A field survey was carried out to determine the relative incidence of CyMV, ORSV and VPV on the different types of symptoms on the Vanilla foliage.

Leaf samples exhibiting symptoms (A) Sunken chlorotic streaks on young leaves and necrotic lesions (<2cm) on older leaves and vines with moderate to severe leaf distortion (Plate 12); (B) Leaf mottles (Plate 13); (C) No Symptom (Plate 14) were collected from 5 plantations with 87 plant samples for A, 61 for B and 105 for C. ELISA was used to test for the presence of CyMV, ORSV and VPV.

Samples were taken from leaf number ten from the shoot tip.

Plate 12. Vanilla leaves showing severe symptom A.
Sunken chlorotic streaks with wavy margin on young *V. fragrans* leaves.

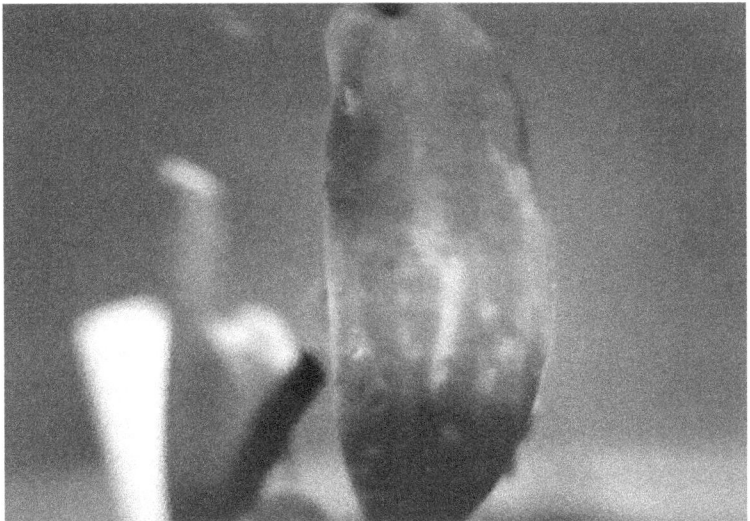

Plate 13. Vanilla leaf showing mottling. Symptom B.

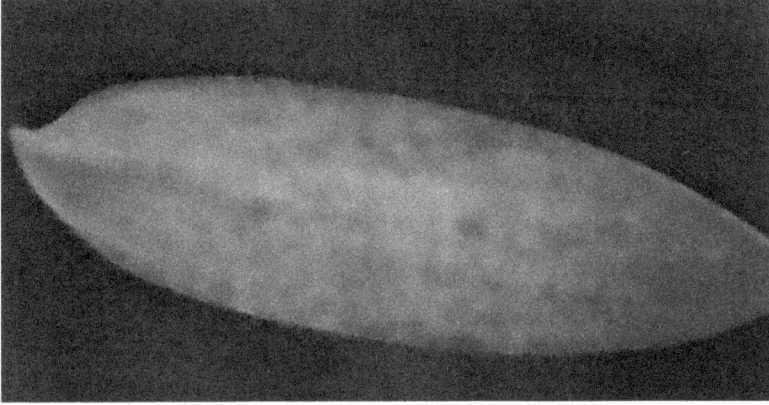

Plate 14. Vanilla leaf with no symptom (C).

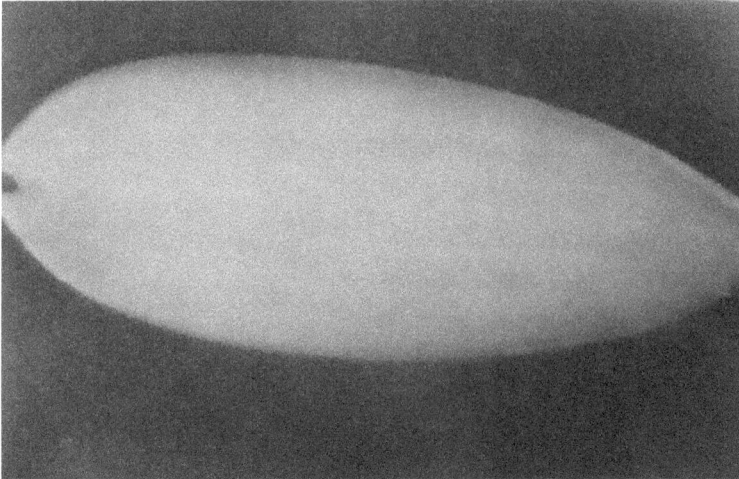

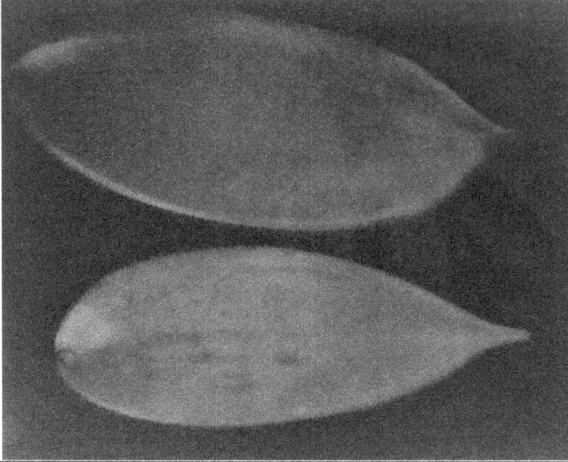

Plate 15. Diffuse, sunken, chlorotic streaks with necrotic spots on young *V. fragrans* leaf (bottom)from VPV inoculated plant. Compare with non-symptomatic leaf at top.

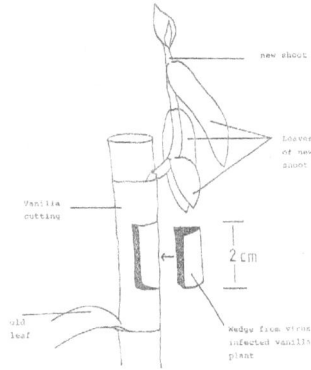

Figure 5.1 Method used to graft virus infected wedges into healthy, young Vanilla plants.

The ELISA test procedure used 8 (Nunc) microtitre wells per plate for the control. Two wells were used per sample. Samples were ground (1:1, w/v) in extraction buffer (Figure 4.3) and incubated at 4°C for 1 hour, for the plant material to settle, before being pipetted into the microtitre wells. A sample was considered positive if the values of the respective wells were greater than the mean plus 2 standard deviations of the control samples.

5.2.2 Pathogenicity tests

5.2.2.1 Grafted plants

All cuttings for these experiments were left in the shade for a week to 'harden' before planting. Four node cuttings from ELISA tested, virus free plants were planted in sterile soil in black polythene bags. The new shoots were tested again at the 2 and 4 leaf stage. ELISA tested virus free cuttings were grafted with tissue wedges taken from surface sterilized pieces of vines infected with CyMV, ORSV, VPV and all three viruses (Figure 5.1). Two sources of infected material were used. The grafts were bound with parafilm and left in an insect screenhouse. Similar grafts were repeated with ELISA tested, virus free wedges as controls. The growth of the plants were measured once a month and foliage observed for symptom development.

5.2.2.2 Mechanically Inoculated Vanilla Plants.

Vanilla cuttings, at the 2-4 leaf stage, were placed in the dark for 3 days. Virus infected and virus free (controls) Vanilla leaves were ground in 0.01 Sorensen's Phosphate Buffer pH 8 (1:1,w/v). The plants were inoculated by lightly sprinkling 600 grit carborundum powder on all leaves, dipping the pestle in inoculum and rubbing the leaf surface with a light pressure [Vanilla leaves have a hard cuticular surface and it was reasoned that a long period in the dark combined with rubbing with a hard surface, rather than a soft surface (pestle vs finger), it might increase the chance of a successful infection]. The inoculated leaves were washed with distilled water after 5 minutes. Plants were put into an insect screen house and their growth rate was measured every month. The foliage was observed for symptom development.

5.2.3 Effect of Virus on Shoot Growth.

The time it took the new shoot to appear and the rate of growth of 4 virus free plants and 4 VPV infected plants were compared over a 4 month period to determine growth rates (Figure 5.2).

5.2.4 Aphid transmission.

Aphis gossypii Glover was raised on healthy *Emilia sonchifolia,* in the laboratory for two weeks. Twenty-five aphids were transferred into a beaker and starved for 90 minutes before they were transferred on to a VPV infected Vanilla plantlet to probe for 30 seconds. Probing was judged by the aphid stopping, the anterior moves down, the antennae moves from the front to the back (Stechmann pers. comm). A magnifying glass was used to observe the probing. After 30 seconds of feeding, the aphid was disturbed with a small paint brush and transferred to the healthy plants. A similar transfer from a healthy to healthy plant was done as the control. Plants were tested for virus infection using ELISA and ISEM after 3 months.

> **Please note:** The number of plants used in this experiment were around 5 to 10 plants for the aphid transmission test and one plant for the control. The author is unsure why the number of plants was not mentioned in the original thesis, probably an oversight.

5.3 RESULTS

5.3.1 Symptoms and virus presence

The incidence of the three viruses singly and in combination in the samples taken are recorded in Table 5.1. The percentage of plants infected by each virus singly are in Table 5.2. Total percentage of plants infected with each virus singly and in combination are recorded

in Table 5.3 and the total incidence of CyMV, ORSV and VPV in the
253* samples taken are given in Table 5.4.

* - There is a figure of 259 in the original thesis which cannot be explained as the
total is only 253, probably an innocent mistake.

Table 5.1. Incidence of single and multiple infections by CyMV, ORSV and VPV in sampled plants from 5 plantations showing foliar symptoms of A, B and C.

Symptom type	A	B	C
Number of plants tested	87	61	105
VPV only	14	2	1
CyMV only	2	2	3
ORSV only	1	15	45
VPV+CyMV	16	1	0
VPV+ORSV	5	1	0
CMV+ORSV	2	24	6
VPV+CyMV+ORSV	42	9	0
No virus detected	5	7	50

Table 5.2 Percentage of plants detected with single infections.

Symptom type	VPV	CyMV	ORSV
A	16	2	1
B	3	3	23
C	1	3	42

93

Table 5.3 Total percentage of plants with single and multiple infections.

Symptom type	VPV	CyMV	ORSV
A	89	71	57
B	26	55	82
C	3	8	50

Table 5.4 Total incidence of viruses (253 samples)

Virus	% Incidence
VPV	37
CyMV	41
ORSV	60

5.3.2 Pathogenicity tests.

The result of the tests, observations and growth measurements of plants grafted with tissue wedges infected with each virus and with all 3 viruses are presented in Table 5.5. Table 5.6 gives the results of the tests and the observations on the mechanically inoculated plants. The presence of the viruses were determined using ELISA and ISEM.

> **NOTE**
> It is worth noting that 5/87 plants (approx. 6%) (Table 5.1), tested with the VNPV antiserum (ELISA test) in this study, were negative or had very weak reactions to be considered positive. It is evidence that there are other 'viruses' causing severe symptoms which were not 'stronglyly' serologically related to VNPV.
> Liefting et al (1992) reported a 'mild' VNPV strain which could protect Vanilla plants against the severe strain.
> Recent developments in the global research on the Vanilla viruses, during the past 31 years, has also reported up to 8 viruses to infect the genus Vanilla including the 3 viruses in this research project.

Table 5.5 Plants mechanically inoculated with ground Vanilla leaves infected with VPV and with all 3 viruses.

Virus presence

	Symptom	VPV	CyMV	ORSV
Control 1	C	-	-	-
Control 2	C	-	-	-
CyMV+ORSV+VPV				
1.	A	+	-	+
2.	A	+	-	+
3.	A	+	+	-
4.	A	+	+	+
VPV only				
1	C	-	-	-
2	A	+	-	-
3	A	+	-	-
Key: A = severe symptom C = no symptom				

5.3.3 Effect of VPV on Vanilla cutting growth.

Figure 5.2 gives the total growth of virus free Vanilla plants compared to VPV infected plants during a 4 month period.

Table 5.6 Plants grafted with tissue wedges infected with CyMV, ORSV and VPV, and by all 3 viruses.

Source		Symptoms	Growth rate (cm)	Virus presence		
				VPV	CyMV	ORSV
Controls						
	1	C	11	–	–	–
	2	C	15	–	–	–
	3	C	15	–	–	–
	4	C	16	–	–	–
VPV only						
Source:Plant						
I	1	A	3	+		
	2	C	21	–		
II	1	C	9	–		
	2	C	17	–		
	3	C	15	–		
	4	A	8	+		
CyMV only						
I	1	C	20		+	
	2	C	22		–	
II	1	C	17		+	
	2	C	20		–	
	3	C	15		+	
ORSV only						
I	1	C	14			+
	2	C	15			–
II	1	C	21			+
	2	C	17			+
	3	C	24			+
VPV + CyMV + ORSV						
I	1	C	18	–	+	+
	2	A	1	+	+	+
	3	A	4	+	+	–
II	1	C	17	–	+	–
	2	C	17	–	–	–
	3	C	22	–	+	+
	4	A	6	+	+	+

KEY: A = severe symptom   C = no symptom

Figure 5.2 Total growth of healthy Vanilla plants (A) compared to VPV infected Vanilla plants (B).

measurement date.	healthy (cm).	std.dev. (cm)	VPV infected (cm).	std.dev. (cm).
26/4	15.3	3.6	1.5	3.
12/5	23.5	7	2.5	3.8
30/5	31.3	9.7	6.3	4.6
27/6	42.5	12	11	0.8
27/7	53	14.6	11	1.6

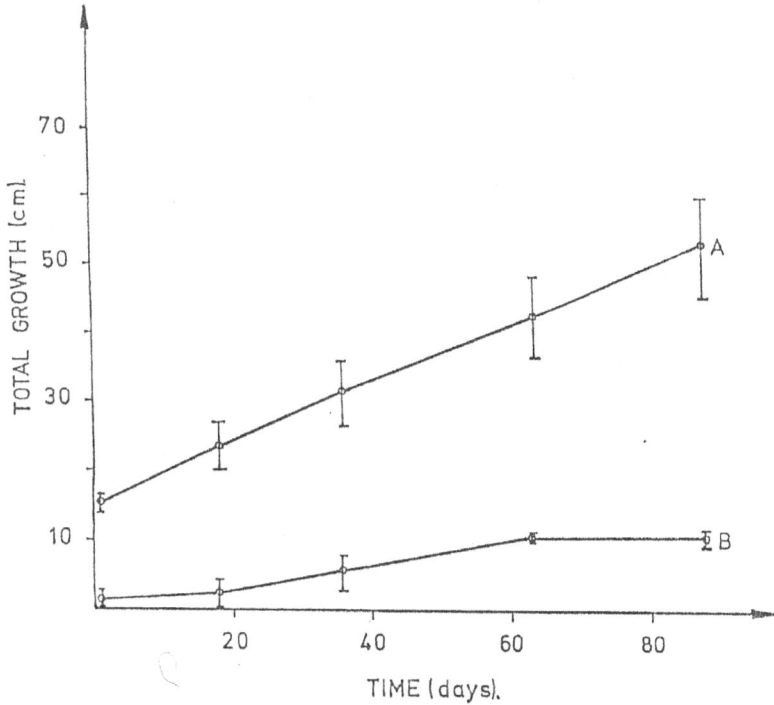

5.4 DISCUSSION

5.4.1 Symptoms and Virus presence.

The results of the field survey (Tables 5.1, 5.2 and 5.3) suggest that VPV causes the severe symptoms described in 5.2.1 (A). In Table 5.3, VPV was detected in 89% of the plants exhibiting severe symptoms compared to 71% and 57% for CyMV and ORSV respectively. VPV was detected in only 26% and 3% in the mottled and non-symptomatic plants. CyMV and ORSV incidences were much higher in these two symptom categories (Table 5.3). It is interesting to note that the presence of the three viruses, in combination, was very high in the severely symptomatic plants (48%). It is therefore worth investigating whether a synergistic effect is occurring. However, field observations could not differentiate any major difference between the occurrence of VPV alone and the presence of all three viruses in the symptoms induced.

ORSV was the most prevalent virus detected in plants exhibiting leaf mottles only. It was also the most prevalent virus in plants exhibiting no symptoms (Table 5.2). CyMV appear to occur mostly in combination with the other two viruses (Table 5.1). The effect of these two viruses need further investigation on a long term basis since their effect may not be readily observed in a short term study.

A difficulty that was realized after the sampling was the shade trees had been pruned prior to the survey in early July.*The survey was probably done in July 1987.* Vanilla plants were therefore exposed to the sun and leaves turn light green which mask any mottling on exposed plants. This would mean that some plants categorized under no symptom might have exhibited mottles under heavier shade. *This may explain why there was such a high ORSV presence (45%) in the 'no symptom' category (Table 5.1).*

98

The high incidence of these three viruses on *Vanilla fragrans* in Tonga is in contrast to results of surveys for viruses on *Vanilla tahitiensis* in French Polynesia. Investigations of the three viruses reported to infect *V. tahitiensis* found 140/1492 (9%) showing potyvirus symptoms, 3/663 (<1%) infected with CyMV and 1/663 (<1%) infected with CyMV and ORSV (Wisler et al, 1987b).

The most commonly detected virus in this Tongan survey was ORSV (60%) followed by CyMV (41%) and VPV (37%) respectively (Table 5.4). The relative incidence of CyMV in this investigation are comparable to results on *Cymbidium* orchids. In a study of the effects of CyMV and ORSV on two cultivars of *Cymbidium*, Pearson and Cole (1986) concluded that ORSV was more infective than CyMV with the infection rates of 70% and 20% respectively. Similarly, Young and Blundell (1979) did a survey of Orchid viruses in New Zealand and found that ORSV was 3 times more common than CyMV. Arunasalam (1987) also reported similar results in a survey of *Cymbidium* in the Auckland area.

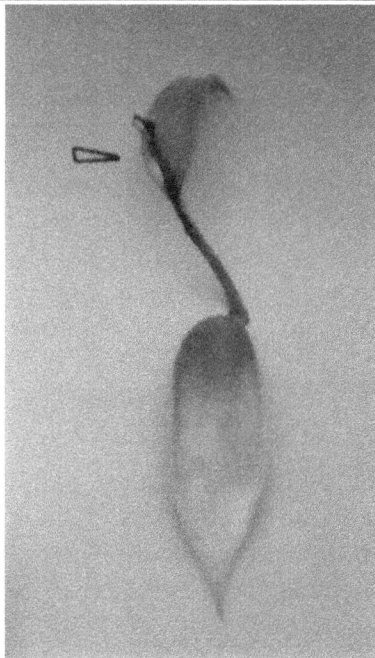

Plate 16. Terminal necrosis (arrow) on a young VPV inoculated *V.fragrans* shoot.

Plate 17. Diffuse sunken chlorosis and terminal necrosis of a VPV infected plant in the field.

However, Wisler *et al* (1982) found that CyMV was more prevalent than ORSV in cultivated orchids in Florida. Wisler *et al* (1987b) also found a higher incidence of CyMV than ORSV in cultivated orchids in French Polynesia. Similarly, Zettler *et al* (1978) reported CyMV to be much more prevalent, in greenhouse raised orchids in Florida, than ORSV.

The difference between the surveys in New Zealand and those in Florida and French Polynesia probably lie in the different sensitivities of the two tests used. ELISA was used for the tests in New Zealand whereas SDS-imunodiffusion was used in the tests in Florida and French Polynesia. As discussed in Section 4.1.1, ELISA is a much more sensitive test than the SDS (sodium dodecyl sulphate) - immunodiffusion test.

5.4.2 Pathogenicity

Although not all the plants inoculated with VPV (Tables 5.5, 5.6) infected material developed symptoms or tested positive to VPV, all the inoculated plants that developed symptoms (11/20 or 55%) of chlorotic streaks , combined with stunted growth were tested positive for the presence of VPV. Some plants that were inoculated with CyMV infected material tested positive to CyMV (3/5 or 60%) but no symptom of effect on growth were observable. Similarly, 4/5 or 80% tested positive to ORSV but no symptom or effect on growth was observable.

Plate 18. Localised, sunken, chlorotic streaks on a young *V. fragrans* leaf from new shoot growth after terminal necrosis.

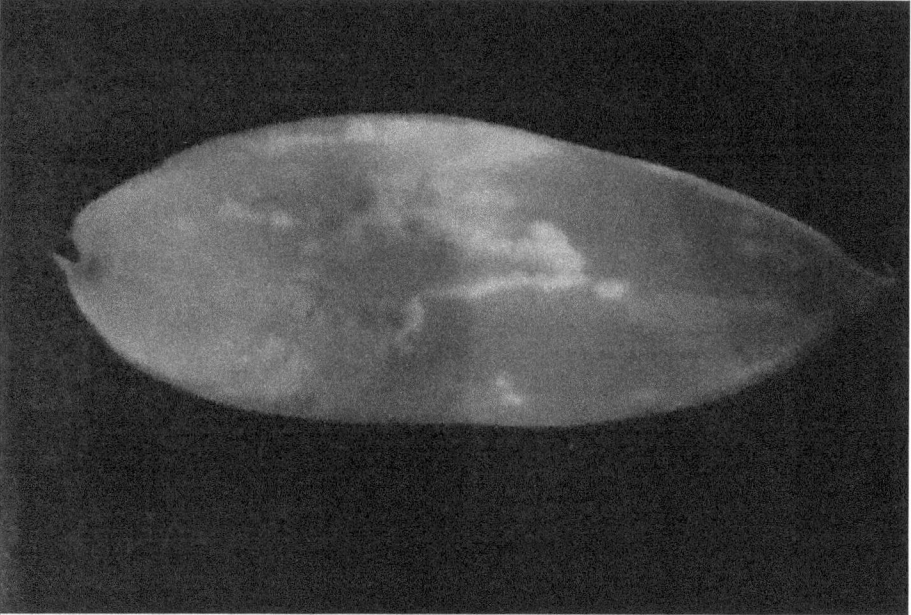

Plate 19. A VPV infected *V.fragrans* in the foreground with most of the leaves gone . This photo was taken after 4 months from terminal necrosis.

The presence of CyMV and ORSV in the multiply infected plants did not appear to have any observable effect on the development of symptoms or the severity of symptoms.

The symptoms that developed on VPV infected plants were diffuse sunken chlorotic streaks and a few necrotic spots (Plate 15). In all these symptomatic plants terminal necrosis followed after the symptoms were observable (Plate 16). Terminal necrosis was also observed in newly infected plants in the field (Plate 17). After about a month, new shoots usually grow from the vines that showed terminal necrosis with the new leaves of such shoots exhibiting leaf distortion and localized chlorotic streaks that sometimes turn necrotic (Plate 18).

In some of the new leaves VPV infected shoots, no apparent symptom was observed. This is especially evident during hot weather. Some potyviruses do cause less severe symptoms at higher temperatures, but cause severe symptoms at lower temperatures (Hollings and Brunt, 1981b). For example, severe red leaf symptoms appeared on certain maize hybrid leaves infected with Maize Dwarf Mosaic Virus when temperatures were lowered from 24-29°C down to 13-18°C for 72 to 96 hours (Hine *et al*, 1970). Similarly, Ross (1970) found that mottling on Pea seed coat caused by Soybean Mosaic Virus had the greatest percentage at 20°C but decreased significantly at 30°C.

The age of the plants inoculated may have also contributed to the recovery of the plants. Thus, Benda and Bennett (1964) showed that when tobacco plants were inoculated with Sugar Beet Curly Top Virus, older plants showed disease symptoms followed by a period of quick recovery. When very young seedlings were inoculated, no clear symptom was observed although the plants contain virus particles. The apparent recovery of the young inoculated Vanilla plants, in the lab, can also be observed in the field to some extent, but most plants in the field will slowly decline and lose most of their leaves after terminal necrosis occurs (Plate 19). This period of slow decline may take 3 months to more than a year. Therefore, the recovery of the young plants, in the laboratory, does not necessarily mean older plants in the field will follow the same pattern. Some terminal necrosis without the sunken chlorotic symptoms of the potyvirus infected plants may also be observed in the field, but usually during dry weather prior to the flowering season. The cause of this terminal necrosis is not known but is thought to be weather induced and therefore initiate flowering naturally. *Growers normally cut or break off the tip of vanilla shoots during the dry season to induce flowering.*

The failure of some of the grafts to transmit virus could be due to low concentration of the virus in the stem wedges used for grafting. Walkey (1985) states that young leaves usually contain more virus than older woody stem. A further reason for this failure could be due to the failure of the vascular tissues to join. Viruses are usually transported, in the plants, quickly through the vascular system (Matthews, 1981). A connection, therefore, between the vascular system of the plant and the infected graft tissue is important for the virus to be transported to the leaves and produce symptoms. However, Muzik (1958) found that vascular tissue connection did not occur between *Vanilla fragrans* and *V. phaentha* Reich. Although some plants were able to survive and grow through parenchyma tissue connections only.

Viruses do not seem to move through parenchyma cells unless they infect the cell and multiply in them, thus resulting in continuous and direct cell to cell invasion. In leaf parenchyma cells, viruses generally move approximately 1mm or 8-10 cells per day (Agrios, 1978). Virus spread between cells is possible through plasmodesmata connections (Gibbs, 1976). It is likely, therefore, that in the Vanilla plants that became infected, in this study, the virus was transported along the parenchyma cells through the plasmodesmata connections and finally got to the vascular tissue where it was transported to the leaves. In plants that failed to produce symptoms, VPV probably did not get to the vascular tissue and therefore could not be transported to the leaves.

The mechanical inoculations were more successful in that only one plant failed to be infected by VPV (Table 5.5). CyMV and ORSV were detected in 2/4 and 3/4 plants, respectively. It is surprising that both CyMV and ORSV were not as infective as VPV in this mechanical inoculation. Infection depends on the successful entry of the virus into a susceptible cell with the help of the abrasive (Walkey, 1985). Thus, the probability for infection depends on the relative concentrations of the virus in the sap. This means that the chance it will come into contact with a susceptible cell will depend on how many viruses are available for infection. Since CyMV and ORSV reach high concentrations in other plants [eg. *Cymbidium* leaves (Francki, 1966) and ORSV in *Nicotiana tabacum cv Samsun* (Paul, 1975)] than potyviruses generally do (Hollings and Brunt, 1981a) it might be assumed that these viruses may be more infective. However, the number of plants mechanically inoculated were too small for any conclusions to be made on relative infectivity.

5.4.3 Effect of VPV on *V. fragrans* growth.

Plate 20. A comparison of 2 healthy and 2 VPV infected V. fragrans plants after 3 months from planting. The tallest healthy plant is about 1 metre high.

The healthy vanilla cuttings took 27-33 days for new shoots to appear whereas the VPV infected plants took 50-113 days. The difference in growth between the VPV infected and the healthy plants (Figure 5.2) indicates that VPV not only delays the new shoot growth, it also stunts overall Vanilla vine growth (Plate 20). Growth reduction in the grafted symptomatic plants (Table 5.6) supports this conclusion. Stunting of growth frequently accompanies observable external symptoms as observed on foliage of VPV infected Vanilla plants.

Latent, symptomless infections also can reduce growth but is more difficult to detect visually (Walkey, 1985). For example, Pearson and Cole (1986) found that CyMV infected Cymbidium had a 100% correlation between symptom expression and CyMV detection by ELISA whereas ORSV did not show any obvious symptoms, yet plants infected by both viruses had a significantly lower mean growth rate than healthy plants. This could mean that CyMV and ORSV can also cause growth reduction in *Vanilla fragrans* in Tonga, but perhaps take a longer period to become apparent.

5.4.4 Aphid transmission

All the plants inoculated by aphid transmission tested negative after 3 months. This could have been due to the small number of aphids used or perhaps the feeding probes were not successful. Wisler *et al* (1987) used 10 aphids to transmit potyvirus infecting *Vanilla tahitiensis* to *V. pompona* Schiede. With only one out of five *V. pompona* plants developing the potyvirus symptom. When 25 aphids per plant were used all the plants infected developed symptoms. Furthermore, the symptoms took 3-8 months to develop, therefore the duration of 3 months for the aphid transmission in this study may not have been long enough, even if the probes were successful.

Field surveys did not find any aphids resident on Vanilla plants, but *Aphis gossypii* and another unidentified aphid were found on herbaceous weeds, *Emilia sonchifolia* and *Sonchus oleraceous*, between Vanilla rows in some plantations. Furthermore, yellow aphid traps set up in one plantation caught a moderate number of the banana aphid, *Pentalonia nigronervosa*, and a few *A. gossypii*. In an effort to find out whether aphids actually alight on the Vanilla plants, insect glue (tanglefoot) was used to cover one leaf on each of 5 plants. The insects trapped were collected after 20 days and identified. Amongst the various insects trapped were 6 winged aphids of the species *Rhopalosiphum maidis* known to feed and colonize various grasses in Tonga. These aphids were observed to probe and feed on Vanilla leaves after starving for one night (Stechmann, 1988 pers.comm.).

The presence of these aphids in Vanilla plantations increases the risk of transmitting VPV from infected to healthy Vanilla plants since it has been shown that *R. maidis* do alight on Vanilla plants and were observed to feed on Vanilla leaves in the laboratory. Pearson *et al* (1990) showed that *Aphis gossypii* and *Myzus persicae* both transmitted VPV from infected to healthy *Nicotiana clevelandii*, but not between infected and healthy vanilla plants.

5.5 CONCLUSION

The results suggest that VPV is the cause of severe symptom (A) on *Vanilla fragrans* foliage and vines. The less than 100% infection in these inoculations may mean that not all sap contact will result in a successful infection.
Vanilla plants infected with CyMV and/or ORSV frequently did not exhibit observable symptoms. However, this may not mean that they cannot cause observable symptoms or retard growth since mottling in the field is only observed in the older leaves and they do stunt the growth of other plants. The plants inoculated were quite young, small in number and observations made over a short duration to detect any effect of these viruses. ORSV was the most prevalent virus in the samples tested with 60% of the samples infected. CyMV and VPV followed with 41% and 37% respectively.

The occurrence of aphids, like *R. maidis*, inside Vanilla plantations, their ability to probe and possibly feed on Vanilla leaves and the transmission of VPV between infected and healthy *N. clevelandii* by *A. gossypii* (*Myzus persicae* was not present in Tonga at the time of this study) poses the threat of VPV being transmitted by these aphids.

However, transmission between infected and healthy Vanilla plants has yet to be demonstrated with any of these aphids.

GLOBAL DEVELOPMENTS SINCE THIS STUDY.

Many other viruses have also been found to infect Vanilla since this study. Grisoni *et al* (2003) mention 7 viruses have been reported to infect Vanilla. 1. Cymbidium Mosaic Virus 2. Odontoglossum Ringspot Virus 3. Vanilla Mosaic Virus (*V.tahitiensis*) 4. Dasheen Mosaic Virus 5. Vanilla Necrosis Virus (VNPV) 6. Cucumber Mosaic Virus 7. Bean Common Mosaic Virus. We can also add a 'mild' strain of VNPV reported by Liefting *et al* (1992) as number 8.

In a study of *Vanilla tahitiensis* in French Polynesia (Grisoni *et al*, 2003) found Bean Common Mosaic Virus, Watermelon Mosaic Virus, Vanilla Mosaic Virus to infect plants in combinations or singly.

Wang *et al*, (1993) suggested that VNPV in this study be renamed WaMV II, after a study of the nucleotide sequence of the protein coat of VNPV suggesting it to be a strain of WMVII, however there remain a large number of answered questions. 1. In Wang's study, some indicator plants showed different symptoms for the two viruses (VNPV and WaMVII) suggesting they are 'biologically' different. 2. Was Wang studying the same virus used for this study? In Table 5.1, there were 5 symptomatic Vanilla plants out of 87 (approx. 6%) which showed the same symptoms but tested negative to the VNPV antibody raised to VNPV and used in the ELISA test suggesting there were a small number of Vanilla plants with the same symptoms but are caused by viruses that are different serologically. Liefting *et al* (1992) also reported a mild strain of VNPV that was different from Vanilla Mosaic Virus and Watermelon Mosaic Virus. In the study by Pearson *et al* (1990), a polyclonal antiserum raised to the Tongan potyvirus failed to react, with the potyvirus associated with mosaic symptoms of *V. tahitiensis* in French Polynesia and ISEM failed to detect any close relationship to 22 other potyviruses suggesting that VNPV is a 'unique' potyvirus. The fact that ELISA Potyvirus Group Tests can be used to test for all potyviruses also suggest that Potyviruses are all 'highly homologus' in a nucleotide sequence study but can be different 'biologically'. In the study by Grisoni et al (2003) in French Polynesia, 'a significant proportion of the plants surveyed (97/476) were symptomatic but tested negative by ELISA for CymMV and the Potyvirus group. Odontoglossum ringspot virus also was not detected in any sample tested.'

Based on the above evidence, it is proposed that renaming the Vanilla Necrosis Potyvirus (in this study) as Watermelon Mosaic Virus II, as proposed by Wang *et al* (1993) may not be scientifically based and it is not supported by ALL the evidence.

....CHAPTER 6....

EPIDEMIOLOGY OF VPV

6.1 INTRODUCTION

6.1.1 General comments on epidemiology

Epidemiology is the study of how a disease is spread through time and space. The source of disease, modes of transmission and the factors affecting them are all important is designing control strategies. Thus, before measures to control a virus disease are formulated and applied it is usually necessary to identify the virus causing it and of vectors spreading the virus, as well as the sources of both virus and vectors (Broadbent, 1964).

VPV was identified as the cause of the severe symptom (A) (Chapter 2). It was then characterized (Chapter 3), partially purified and an antiserum was developed (Chapter 4) because a test was required to identify this virus in the field. The pathogenicity of VPV on Vanilla plants was determined by inoculating young Vanilla plants with VPV (Chapter 5). The symptoms produced appear to be the same with those observed in the field (Plates 15, 16, 17). Furthermore, 89% of symptom A plants surveyed were tested positive to VPV using ELISA (Table 5.3). Therefore, it is possible to use the occurrence of symptoms as an indicator of VPV presence in Vanilla plants in the field.

The results in the previous chapters suggest that VPV is the virus causing the most damage to Vanilla plants and plantations. Since no apparent effect of CyMV and ORSV were found apart from a possible cause of mottling and time restrictions cannot allow for extended research at this stage, it was decided that for the purposes of an immediate short term control programme, the epidemiology of VPV should be investigated first. The results of these short term control measures will then be considered in designing long term control measures (See Leaflet in Appendix 6).

Aphid transmitted viruses can be carried into a crop by aphids after feeding on infected nearby plants. However, the duration of time between the last probe and the next one is important. For non-persistent viruses like the potyviruses, the aphid loses the ability to transmit these viruses after a short time (Walkey, 1985). Therefore, aphid transmission of these viruses may be restricted to short distance transmission within crops or between adjacent plantings.

The distribution of virus infected individuals in a plant community, such as a crop, can provide useful clues to the sources and mode of spread of a virus (Gibbs, 1983). Therefore, the spatial distribution of VPV infected plants in Vanilla plantations can be used to make preliminary conclusions on how the virus is spread within a plantation. In addition, the wider distribution and spread of viruses between plantations and between islands was considered in relation to the use of infected planting material and possible insect transmission.

6.1.2 General distribution of VPV in Tonga.

The allocation of funds and manpower for the control of a disease depends on how widespread the problem is. The larger the area affected, the more money and manpower required for virus control, since most virus control methods with vegetatively propagated plants will largely depend on removal of infected plants. A localised disease problem may not warrant the same concern as a widespread one. Therefore, a survey of the dispersal of the virus is necessary.

6.1.3 Source of Infection.

The vegetatively derived progeny of systemically virus infected plants are also virus infected. Such infected planting material use to establish a new crop are a potent source of virus for the crop (Gibbs and Harrison, 1976).

This will cause the problem that all the future progeny of these plants will also be virus infected. The infection will probably result in a reduction of yield and quality, while the cuttings can be a source of virus infection for other crops. Infected planting material also provide a means by which a virus may spread internationally over great distances. For example, the spread of Plum Pox Virus in Europe and Grapevine Fan Leaf Virus throughout the world, in infected rootstocks, are classical examples of this dispersal (Walkey, 1985).

Vanilla fragrans is a vegetatively propagated plant. In Tonga, new Vanilla growers normally buy their cuttings from established growers resulting in a trade of Vanilla cuttings between the islands of Vava'u, where most of the Vanilla is grown, and the rest of the Tongan Islands especially the main island of Tongatapu.

Since the symptoms of VPV are readily observed on the Vanilla foliage, any cutting taken from an infected plant will show the symptoms described in Section 5.2.1 (Plates 12, 15, 17, 18) on its old leaves. The new shoots germinating from such cuttings will also show the symptoms. Thus, in new plantations (<2 years old) it is possible to determine whether the infection is from the initial cutting and the new shoot by observation only.

6.1.4 Two dimensional distribution of VPV infected plants in Vanilla plots.

The pattern of distribution of virus infected plants in the field can be used to study the spread of virus infections and therefore become useful in designing control strategies. The most convenient method to study the spread of a disease is to study the distribution of diseased plants in the field or orchard without the knowledge of their history. On average, diseased plants are often found to clump around other formerly infected plants (van der Plank, 1960).

Viruses do not spread on their own, but are carried by vectors. The vector for potyviruses are usually aphids transmitting the virus in a non-persistent manner (Hollings and Brunt, 1981a; Matthews, 1982). VPV is known to be transmitted in a non-persistent manner by *Aphis gossypii* and *Myzus persicae* (Pearson *et al*, 1990). Various other aphids like *P.nigronervosa* and *R.maidis*, have been trapped in Vanilla plantations and Vanilla leaves (Section 5.4.4). Potyviruses can be transmitted after a brief probe of several seconds (Walkey, 1985) and therefore the presence of these aphids inside Vanilla plantations and Vanilla leaves will increase the chance of VPV being transmitted to healthy plants.

The spatial pattern of infected plants will be clustered or clumped if the infection of a given plant increases the probability of another plant nearby, being infected. In short, with clustering, infected plants are, on average, grouped together and healthy plants are grouped. The method for determining the spatial pattern depends on collecting several independent samples or dividing a field into contiguous quadrats (Figure 6.1, 6.2 and 6.3) (Madden and Campbell, 1986).

The pattern of distribution of the population of VPV infected Vanilla plants can then be determined by comparing the variance (s^2) and the arithmetic mean (u) of the quadrats in each plot. If $s^2 < u$ then the distribution is a positive binomial or regular. If $s^2 = u$ the distribution is that of a poisson series or random. However, when $s^2 > u$ the distribution is of a negative binomial or contagious (clustered) (Elliot, 1977).

A test for significance in this case is given by;

$x^2 = s^2 (n-1)/u$

When the x^2 value is greater than the corresponding value in a chi square table with n-1 degrees of freedom and significance level P, one rejects the poisson distribution in favor of clustering (Madden and Campbell, 1986).

6.1.5 Linear distribution of VPV infected plants in Vanilla plantings.

6.1.5.1 Distribution of VPV infected plants along Vanilla rows.

Mechanical transmission of viruses in the field by natural mechanical damage to the plant tissue is relatively rare, and probably of very minor economic importance. A more common means of mechanical transmission in the field is through normal horticultural practices. For example, Tobacco Mosaic Virus maybe transmitted in tomato crops by contaminated hands, clothing and tools. Many other viruses may be transmitted by unsterilized tools during pruning procedures and when cuttings are taken (Walkey, 1985).

Vanilla fragrans are planted in rows with a spacing of 1.5 m between plants and 2.5 m between rows giving rise to approximately 2200-2500 plants per hectare. Vanilla growers normally work along the rows (Section 1.3.1). Since VPV can be mechanically transmitted (Section 5.3.2), it is highly likely that it is spread along the rows during cultural practices such as looping, flower initiation and taking of cuttings. If this assumption is true, then the actual number of consecutively infected plants along a row will be greater than the number of consecutive infected plants expected by a chance infection.

Therefore, a test for the spread of virus infection is a test for aggregations of virus infected plants along a row (van der Planck, 1960). Van der Planck (1960) described a test for such aggregations based on the number of adjacent infected plants called a 'doublet analysis'. A doublet consist of 2 adjacent infected plants; 3 adjacent infected plants is considered as 2 doublets; 4 adjacent infected plants is considered as 3 doublets and so on. The Null Hypothesis evaluated is that the ordered sequence is random; the alternative hypothesis is the sequence is clustered (Madden and Campbell, 1986).

If **n** plants are investigated in a sequence, and of these **u** are diseased, then the expected number of doublets **d** is given by;

d = u(u-1)/n

(van der Planck, 1960).

A better form of this equation was given by Madden and Campbell (1986) as;

E(D)   m(m-1)/N

Where **E(D)** is the expected number of doublets, **m** is the number of infected plants in a row and **N** is the total number of plants in that row.

The standard deviation **s(D) = {[m(m-1)/N](1-2)/N)}** $^{1/2}$

For rows with **N>20** a test of randomness is given by;

Z=(D-E(D))/s(D)

If **Z>1.64**, one rejects the Null Hypothesis of randomness in favor of clustering **(P=0.05)**.

A second method with the same assumptions was proposed Madden et al (1982). They proposed an 'ordinary run analysis' where in sequences of diseased and healthy plants in a row, a run is defined as a consecutive clump of either healthy or infected plants. For example, if the signs - and + stand for healthy and infected plants respectively, then in the sequence ++-+++----+, runs will be '++', '-', '+++', '-----', and '+'. Therefore this sequence contain **5** runs. Using this system, the expected number of runs **E(U)** is given by;

$$E(U) = 1+2m(N-m)/N$$

With standard deviation;

$$s(U) = \{(2m(N-m)[(2m(N-m)-N]/[N^2(N-1)]\}^{1/2}$$

A test for randomness **(N>20)** is given by;

$$Z(U) = [U+0.5-E(D)]/s(U)$$

If **-Z(U) >1.64** then one rejects the Null Hypothesis of randomness in favor of clustering.

6.1.6 Spread of VPV with time.

The infection of isolated plants by viruses is not normally of economic importance. Viruses become important when they cause epidemics. Epidemics may be defined by a change of disease intensity with time. Disease intensity has two components (1) disease incidence, which is the proportion of infected plants (2) disease severity, which is the number of plants infected (Madden and Campbell, 1986). The increase of disease is usually from a focal point, which is formerly infected plant (Gibbs, 1983). Therefore, a visual diagram of the spread of the virus through time is essential for a clear understanding of the problem (Figure 6.4).

6.2 MATERIALS AND METHODS

The diagnosis of a VPV infected plant in these studies are based on the presence of the symptoms attributed to VPV (Chapter 5). The presence of severe symptoms is highly correlated to the presence of VPV. Therefore, no tests to confirm the presence of VPV were done due to the large numbers of plants involved.

6.2.1 General distribution, source of infection and spread of VPV in Tonga.

Two hundred and thirteen Vanilla plantations on the islands of Vava'u and Tongatapu were randomly selected from the Ministry of Agriculture, Fisheries and Forest's list of Vanilla Growers using the first three digits of a random numbers table. Of these, only 45 were finally surveyed due to the large number of neglected and abandoned plantations (Appendices 3 and 4). For plantings under 2 years old, the original cuttings were examined to determine whether symptomatic shoots grew from infected cuttings. Vanilla plantings more than 2 years old were also surveyed to determine presence or absence of VPV symptoms on plants. Every fifth row was examined and the number of healthy and infected plants were recorded.

Figure 6.1. Map of healthy and VPV symptomatic vanilla plants in Plot 1.

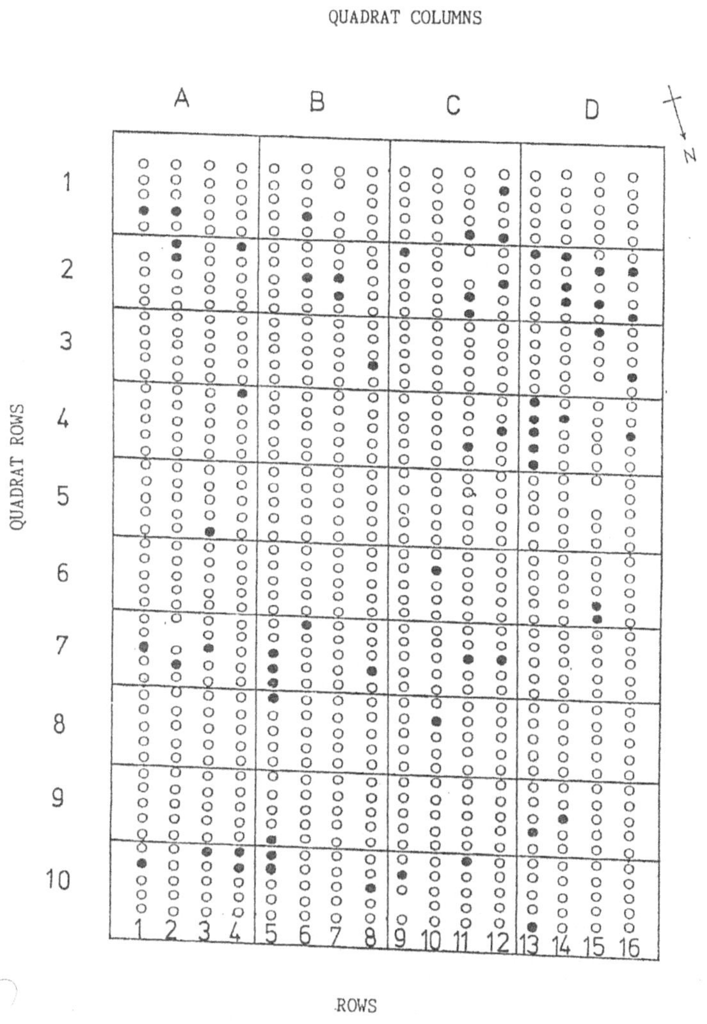

Figure 6.2 Map of healthy and VPV symptomatic vanilla plants in Plots 2.

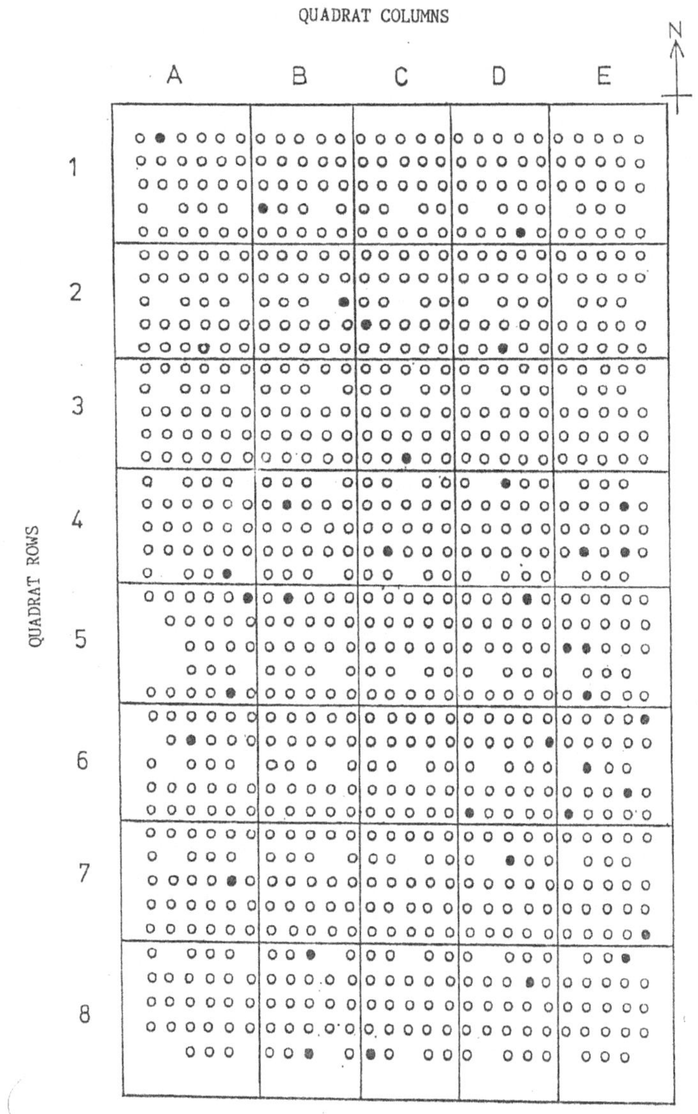

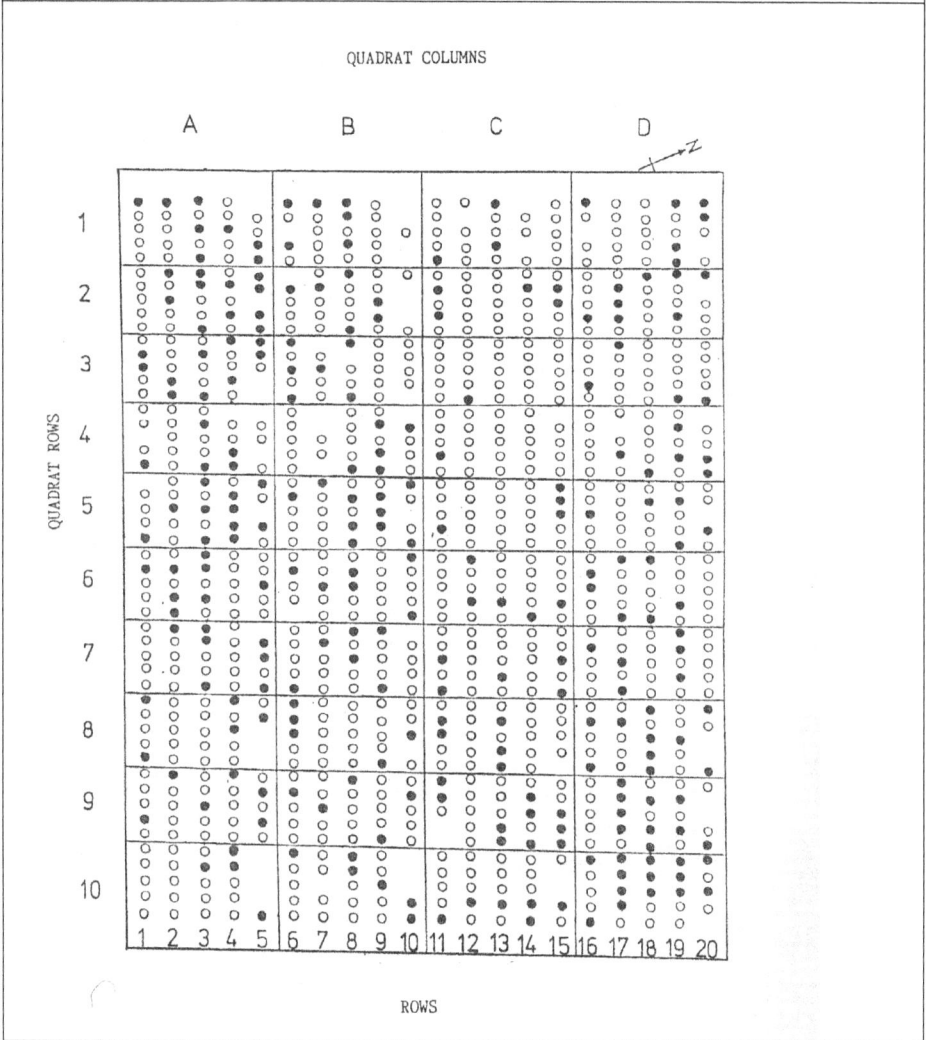

Figure 6.3 Map of healthy and VPV symptomatic vanilla plants in Plot 3.

6.2.2 Spacial distribution of VPV infected plants in the Vanilla plantings examined.

Three Vanilla plantations, two from Tongatapu (Figure 6.1 and 6.2) and one from Vava'u (Figure 6.3), were mapped by marking all plants in every row and distinguishing between VPV infected and non-symptomatic plants. Plots 1 and 2, on Tongatapu, were monitored every month for 5 months to provide information about the spread of VPV through time.

The results of the mapping were then analysed for contagious spread in the two dimensional distribution of infected plants inside the Vanilla plantings using the method of Elliott (1977) reviewed by Madden and Campbell (1986) (Section 6.1.4). Each plot was divided into contiguous quadrats of 20 plants (Plot 1), 24-30 plants (Plot 2) and 21-25 plants (Plot 3) (Figures 6.1, 6.2, 6.3) and the mean and variance of the number of VPV infected plants were compared. The relationship between the mean and variance indicates the nature of the distribution. The same procedure was repeated using larger quadrats at twice the size of the original quadrats, for each plot, for comparison (Table 6.6). The chi-squared test was then used to test for any significant deviations from the Poisson Distribution. Results are shown in Table 6.7.

Clustering of VPV spread along the rows of Vanilla plants was also analysed using the 'doublet' analysis of van der Planck (1946) reviewed by Madden and Campbell (1986) and the 'ordinary run' analysis of Madden et al (1982) (Section 6.1.5). The results are presented in Tables 6.8, 6.9, 6.10 and 6.11.

6.3 RESULTS

6.3.1 General distribution of VPV in Tonga

A total of 39 plantations, over two years old, were surveyed for VPV symptoms. The proportion of symptomatic plants in the samples surveyed varied from 0-83%, with an average of 12%. No symptomatic plants were found in 4 of the plantations sampled and 35 plantations had symptomatic plants in the rows surveyed. In 32 plantations, the proportion of symptomatic Vanilla plants, in surveyed rows, was less than 30%. Only 3 sampled plantations had very high incidences of symptomatic plants, V5 (35%), V15 (56.7%) and V18 (83%). The complete results are presented in Table 6.1.

6.3.2 Source of infection

Eight new plantations, less than 2 years old, were surveyed for the presence of symptomatic shoots with symptomatic cuttings. The proportion of symptomatic shoots growing from symptomatic cuttings was 100%. All the symptomatic shoots, were observed, to grow from symptomatic cuttings. The complete results are presented in Table 6.2.

6.3.3 The pattern of distribution of VPV infected plants in Vanilla plantings.

6.3.3.1 Two dimensional distribution of VPV in Vanilla Plots 1, 2 and 3.

The comparison of the mean and the variance in the three plots monitored found the distribution of the infected plants to follow a negative binomial or contagious distribution. The variance was greater than the mean number of infected plants in the quadrats of each plot. For the smaller quadrat size, the test for significant deviations from the Poisson Distribution found Plot 1 to differ significantly (P=0.001) whereas Plots 2 and 3 were not.

When the size of the quadrats were increased to twice the original size, all the plots were found to differ significantly from the Poisson Distribution at P=0.10. Only Plots 1 and 3 differed significantly, from the Poisson Distribution at P=0.05 and P=0.025 respectively. None of the plots differed from the Poisson Distribution, at P=0.01, when the quadrat size was doubled (Table 6.7).

6.3.3.2 Linear distribution of VPV infected plants in Vanilla rows.

Plot 1 had significant clustering in rows 5,7 and 13 in both the doublet and the ordinary run analyses. Although Plot 3 appear to have clustering of VPV infected plants along the rows, by visual inspection, no significant clustering was found by the doublet analysis or the ordinary run analysis. It is worth noting that there were some high positive values in the doublet analysis (rows 14, 15 and 18). These were confirmed by corresponding high negative values in the ordinary run analysis. These values may mean that these rows had a significant deviation from the Poisson Distribution, but at lower levels than required by the test.

Plot 2 was not analysed using this method as visual inspection does not show any consecutive VPV symptomatic plants (doublets) along the rows. The complete results are presented in Tables 6.7, 6.8, 6.9 and 6.10.

6.3.4 Spread of VPV with time

The increase in the number of symptomatic plants was quite rapid in Plot 1 with an increase of 40 plants (160%) exhibiting symptoms in 4 months. Plot 2 had only two newly infected plants during the same period. The increase in the number of symptomatic plants observed in Plot 1 coincided with a period when a large number of aphids were colonizing some herbaceous weeds inside this plot. No aphids were found in plot 2 during the monitoring period, as it was always 'weeded clean' by the grower (Table 6.12).

Table 6.1 Proportion VPV infected plants in samples surveyed from plantations over 2 years old.

Plot number.	sample number.	number of symptomatic plants.	percentage of symptomatic plants.
V2	111	16	14
V3	94	0	0
V4	87	0	0
V5	74	26	35
V6	87	7	8
V7	164	36	22
V8	321	49	15
V9	148	8	5
V11	87	15	17
V12	91	7	8
V13	142	0	0
V14	252	16	6
V15	97	55	57
V16	162	5	3
V17	112	5	4
V18	130	108	83
V19	124	6	5
V20	173	48	28
V21	127	14	11
V22	99	2	2
V23	83	5	6
T1	160	21	13
T2	129	1	1
T4	109	5	5
T5	152	2	1
T6	140	7	5
T7	136	5	4
T8	90	1	1
T9	160	10	6
T10	75	6	8
T11	80	1	1
T12	147	5	3
T13	281	6	2
T14	24	5	21
T15	115	24	21
T16	55	0	0
T19	114	7	6
T21	147	12	8
T23	85	23	27

Tables 6.2 and 6.3.

Table 6.2 Number of symptomatic shoots (< 2 years old) with symptomatic cuttings.

		Number of plants exhibiting VPV symptoms.	
Plot number.	sample number.	new shoot.	original cutting.
T3	264	17	17
T17	55	4	4
T18	97	0	0
T20	137	1	1
V1	106	4	4
V10	212	32	32
V29	45	0	0
V30*	592	295	295

* All plants in this plot were recorded.

Table 6.3 The number of symptomatic vanilla plants per quadrat in plot 1.

		quadrat columns			
		A	B	C	D
	1	2	1	3	0
	2	3	3	4	8
	3	0	1	0	2
	4	1	0	2	7
quadrat	5	1	0	0	0
rows	6	0	0	1	2
	7	3	5	2	0
	8	0	1	1	0
	9	0	1	0	2
	10	4	3	2	1

Tables 6.4 and 6.5.

Table 6.4 The number of symptomatic vanilla plants per quadrat in plot 2.

		quadrat columns			
		A	B	C	D
	1	1	1	0	1
	2	0	1	1	1
	3	0	0	1	0
quadrat	4	1	1	1	1
rows	5	2	1	0	1
	6	1	0	0	2
	7	1	0	0	1
	8	0	2	1	1

Table 6.5 The number of symptomatic vanilla plants per quadrat in plot 3.

		quadrat columns			
		A	B	C	D
	1	8	6	3	6
	2	11	6	5	8
	3	10	6	1	4
	4	5	5	1	6
quadrat	5	11	10	4	5
rows	6	8	6	5	7
	7	7	6	5	6
	8	5	5	5	10
	9	6	5	10	10
	10	4	6	6	14

Table 6.6 and 6.7.

Table 6.6　The variances and the means of plots 1, 2, and 3, for the two quadrat sizes used.

| | | quadrat size | | |
| | large | | small | |
plot	mean	variance	mean	variance
1	3.3	6.5	1.675	3.76
2	1.8	2.7	0.875	0.8814
3	12.8	22.48	6.45	7.06

Table 6.7　The chi-squared statistic for plot 1, 2, and 3, for the two quadrat sizes used at different levels of significance.

		quadrat size				
Plot	large			small		
	x^2 cal.	x^2 tab.(19 d.f.).		x^2 cal.	x^2 tab.(39 d.f.).	
		P = 0.10　0.05　0.025			P = 0.10　0.05　0.001	
		27.2　30.14　32.85			51.8　55.7　73.4	
1	37.42			87.54		
2	28.5			39.28		
3	33.37			42.7		

Tables 6.8 and 6.9.

Table 6.8 Doublet analysis of the distribution of VPV infected plants along vanilla rows in plot 1.

Rows	D	m	N	E(D)	s(D)	Z	
1	0	3	49	0.122	0.342	−0.356	
2	1	4	49	0.244	0.48	1.57	
3	0	3	50	0.12	0.339	−0.353	
4	1	4	50	0.24	0.48	1.58	
5	5	7	49	0.857	0.9	4.6	*
6	0	3	50	0.12	0.339	−0.353	
7	1	2	50	0.04	0.196	4.89	*
8	0	3	50	0.12	0.339	−0.353	
9	0	2	49	0.04	0.196	−0.558	
10	0	2	50	0.04	0.196	−0.558	
11	1	6	49	0.6122	0.766	0.506	
12	0	5	50	0.4	0.619	−0.646	
13	4	8	50	1.12	1.037	2.77	*
14	1	5	50	0.4	0.619	0.969	
15	1	5	48	0.416	0.63	0.926	
16	0	4	50	0.24	0.48	−0.5	

* − significant at P = 0.05

Table 6.9 Ordinary run analysis of the distribution of VPV infected plants along rows of vanilla in plot 1.

Rows	U	m	N	E(U)	s(U)	Z	
1	7	3	49	6.6	0.736	1.22	
2	7	4	49	8.3	0.985	−0.812	
3	7	3	50	6.64	0.956	0.899	
4	7	4	50	8.36	0.977	−0.880	
5	5	7	50	13.04	1.64	−4.59	*
6	7	3	50	6.64	0.73	1.17	
7	3	2	49	4.83	0.47	−2.829	*
8	7	3	50	6.64	0.73	1.17	
9	5	2	49	4.8	0.475	1.47	
10	5	2	50	4.84	0.471	1.40	
11	11	6	49	11.5	1.44	0	
12	11	5	50	10	1.212	1.237	
13	9	8	50	14.44	1.846	−2.676	*
14	9	5	50	10	1.21	−0.413	
15	9	5	48	9.9	1.23	−0.325	
16	9	4	50	8.36	0.448	2.54	

* − significant at P = 0.05

Table 6.10 and 6.11

Table 6.10   Doublet analysis of the distribution of VPV infected plants along vanilla rows in plot 3.

Row	D	m	N	E(D)	s(D)	Z
1	1	9	49	1.46	1.18	-0.389
2	3	11	50	2.2	1.45	0.55
3	6	21	50	8.4	2.83	-0.848
4	4	16	49	4.89	2.21	-0.4
5	7	18	35	8.74	2.87	-0.606
6	3	14	48	3.79	1.9	-0.415
7	0	7	45	0.93	0.89	-1.04
8	5	18	49	6.2	2.44	-0.491
9	4	13	50	3.12	1.73	0.51
10	2	9	35	2.05	1.39	-0.036
11	2	12	48	2.75	1.62	-0.46
12	0	4	49	0.24	0.48	-0.5
13	2	10	50	1.8	1.31	0.625
14	2	7	48	0.875	0.916	1.228
15	5	12	47	2.8	1.64	1.34
16	1	11	49	2.2	1.468	-0.817
17	8	18	49	6.2	2.44	0.737
18	6	15	50	4.2	2	0.9
19	5	20	50	7.6	2.7	-0.96
20	3	12	41	32	1.75	-0.114

Table 6.11   Ordinary run analysis of the distribution of VPV infected plants along vanilla rows in plot 3.

Rows	U	m	N	E(U)	s(U)	Z
1	16	9	49	15.7	2.04	0.392
2	16	11	50	18.16	2.37	-0.7
3	30	21	50	25.36	3.4	1.5
4	23	16	49	22.55	3.03	0.31
5	20	18	35	18.4	2.9	0.724
6	22	14	48	20.8	2.81	0.6
7	12	7	45	12.8	1.7	-0.17
8	26	18	49	23.7	3.21	0.872
9	19	13	50	20.24	2.67	-0.277
10	14	9	35	14.37	2.2	0.059
11	20	12	48	19	2.55	0.588
12	9	4	49	7.3	0.98	2.24
13	16	10	50	16	2.21	0.226
14	10	7	48	11.95	1.67	-0.868
15	15	12	47	17.87	2.56	-0.925
16	19	11	49	18.06	3.03	0.475
17	21	18	49	23.77	3.21	-0.7
18	19	15	50	22	2.92	-0.856
19	30	12	50	24	3.35	1.94
20	18	12	41	17.97	2.6	0.2

Table 6.12 The number of VPV symptomatic plants in Plots 1 and 2 per month, between August and December 1987.

Table 6.12   The number of VPV symptomatic vanilla plants in plots 1 and 2 per month, between August and December (1987).

Month.	plot 1.	plot 2.
August	25	34
September	25	34
October	36	34
November	45	35
December	66	36

NOTE

It is worth noting that the rapid increase of infected Vanilla plants in Plot 1 from 25 plants to 66 plants occurred from August to December. A time of increasing warm temperatures in Tonga where August is 'cool' and December is fairly 'hot'. The warmer temperatures may have induced the large aphid colonies on weeds inside Plot 1 to breed and develop wings then migrate. This migration also include landing and probing on plants thus spreading the potyvirus from infected to healthy Vanilla plants. Also note that Plot 2 was 'weeded clean' by the grower every week and only had 2 new infections during the same period.

Figure 6.4. Spread of VPV in Plot 1 from September to December, 1987.

Figure 6.4   Spread of VPV in plot 1 from September to December, 1987.

[Key: 0 - symptomatic plants in September; 1 - symptomatic plants in October; 2 - symptomatic plants in November 3 - symptomatic plants in December].

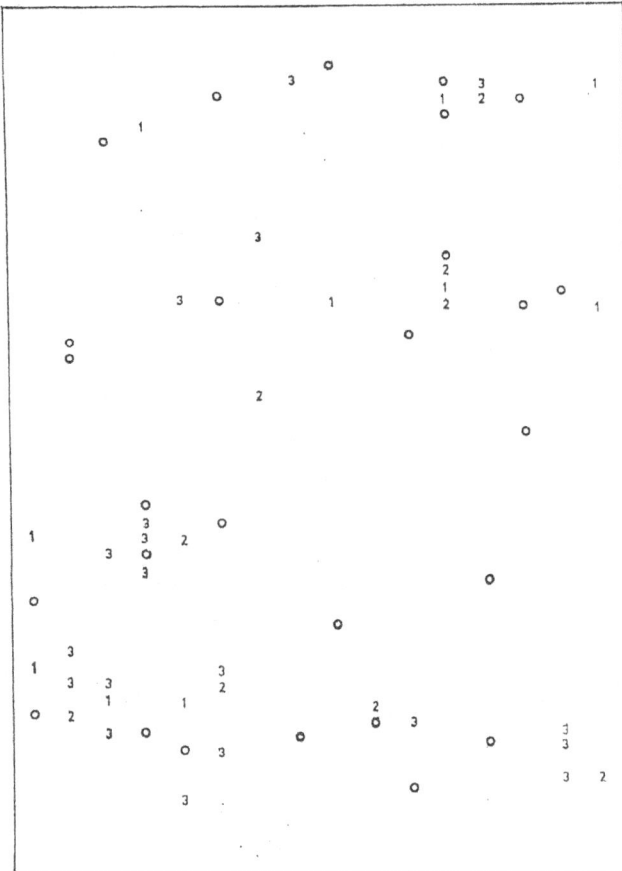

130

6.4 DISCUSSION

6.4.1 General distribution of VPV in Tonga.

The survey of 39 plantations on the islands of Vava'u and Tongatapu showed the VPV infected plants were widespread on these islands (Table 6.1, Appendices 3 and 4). Most of the plantations surveyed (82%) appear to have a low to moderate incidence of VPV symptomatic plants (<30%) with a few plantations with high incidence of VPV (>30%).

The age of the plantations surveyed was between 3 and 8 years. The variability of the incidence of symptomatic plants between plantations of different age means that VPV may not be spreading very quickly. If VPV had spread very quickly within plantations, then one would expect a significant trend of increasing VPV incidence with the age of the plantations. However, there may be isolated cases of rapid VPV spread like those plots V5, V15 and V18 (Table 6.1). A difficulty in the interpretations of the data is the fact that VPV infected plants have a variable longevity. Some plants seem to die off quickly, within 3 months, while others appear to survive for a year or more. The survey would have been looking at the incidence of the symptomatic plants, at one point in time, with a tendency to pick up long lived symptomatic plants. A continuous survey over a longer period would be better; but not possible in the short duration (2 years) of this study.

6.4.2 Source of infection.

The results in Table 6.2 showed that the plants showing symptoms originated from VPV infected cuttings, therefore there is a 100% correlation between infected shoots and cuttings in young Vanilla plantings less than 2 years old. Inspection of some cuttings, from Vava'u, to be planted on Tongatapu found some of the cuttings to exhibit symptoms common to VPV infected plants. It is well known that the use of systemically infected cuttings, in the multiplication of a vegetatively propagated horticultural crop like Vanilla, will result in

131

the spread of the viruses through the cuttings. The progeny of these infected plants will also be infected. Therefore, from the evidence, it is highly likely that VPV infected Vanilla plants in new plantings originate from infected plants.

Although the incidence of infected cuttings was quite low, the high incidence of symptomatic cuttings in Plot V30 (Table 6.2) demonstrates what could happen if growers do not ensure that only healthy, vigorous, non-symptomatic cuttings are used as propagating material. The number of dead Vanilla plants were not counted, but if death were due to virus infection then the true incidence of the VPV disease is greater than recorded.

6.4.3 The two dimensional distribution of VPV infected plants in Vanilla Plots 1, 2 and 3.

The pattern of distribution of VPV infected plants in all the Vanilla plots studied fits the negative binomial model (Table 6.6). This means that the distribution is contagious, so the presence of a diseased plant increases the probability of nearby healthy plants getting infected. The distribution of VPV symptomatic plants in these plots are therefore clumped.

Virus transmission by aerial vectors often result in diseased plants being in expanding patches or clumps (Gibbs and Harrison, 1976). Since potyviruses are known to be transmitted mostly by aphids, in a non-persistent manner, and through mechanical transmission (Hollings and Brunt, 1981a; Matthews, 19820, the two dimensional pattern of symptomatic plants in these three plots is most likely to be the result of VPV transmission by winged aphids.

Although no aphid have been found to colonize Vanilla plants in Tonga, it is known that aphids do frequent Vanilla plantations, and also colonize herbaceous weeds inside plantations. It is also known that aphids do alight on Vanilla plants and also probe or feed on Vanilla leaves in the laboratory (Section 5.4.4). Mechanical transmissions in 2

132

dimensions is highly unlikely since Vanilla growers normally work along the Vanilla rows (Section 6.3.3.2).

The differences in the results for the two quadrat sizes could mean that the size of the quadrat affects the final conclusion. Elliott (1977) suggested that there are different types of contagious distributions. For example (i) small clumps isolated from each other (ii) large clumps with individuals distributed randomly in each clump (iii) large clumps with individuals distributed uniformly in each clump. Therefore, for a distribution with small clumps of infected plants, isolated from each other, a small sized quadrat maybe required to find any significant deviation from the Poisson Distribution. This may be the case in Plot 1 where decreasing the quadrat size had increased the sensitivity of the test from $P=0.025$ to $P=0.001$.

In a distribution where there are large clumps of infected plants with individuals randomly distributed in each clump, a small quadrat may fail to pick up the clumping and ultimately a conclusion of random distribution may be reached. This maybe the case in Plot 3 where there were a large number of VPV infected plants. The larger sized quadrats resulted in conclusions of significant deviations from the Poisson Distribution at $P=0.025$, whereas the smaller quadrat size showed results of no significant deviation from the Poisson Distribution, even at $P=0.10$. It seems necessary, therefore, that any investigation of this type make use of different quadrat sizes to get a complete picture of the kind of distribution in the plantations in the study.

6.4.4 Linear distribution of VPV infected plants in Plots 1 and 3.

The possibility of VPV being transmitted by aphids in the Vanilla Plots, does not lessen the likelihood of VPV spreading along Vanilla rows through mechanical transmission during looping, flower initiation (Section 1.3.1) and the taking of cuttings. Therefore, the analysis of spread along Vanilla rows, although superficial, is still worthwhile.

The three rows in Plot 1 (Tables 6.6, 6.7), that showed a significant clustering of diseased plants demonstrated the possible effects of spreading VPV along the Vanilla rows. Although Plot 3 had a higher incidence of VPV infected plants, no significant effect of VPV along the rows was detected.

Virus infection of plants normally take time to show symptoms. The period between infection and expression of symptom vary between different viruses and is known as the latency period (Bos, 1978). It was difficult, was difficult to know the time of infection or tell whether a plant is infected or not if no symptoms are apparent. Hence, the use of symptoms as an indicator of VPV presence may underestimate the incidence of VPV infection, although in our survey in Chapter 5 (Table 5.1), less than 1% of 105 non-symptomatic Vanilla plants tested positive to VPV.

Although the two methods of analysis used here seem to work quite well, they rely on van der Planck's postulate that 'clusters of infected plants resulted from a pathogen spreading from plant to plant' (Madden et al, 1982). This postulate appear to be acceptable. The disadvantage with the test is that they do not show the origin of the infection, whether it came from within the plantation or from outside the plantation. Madden et al (1982) states that the 'ordinary run analysis' is a better method for testing clustering or randomness of infected plants than the 'doublet analysis' method. *They concluded that the doublet analysis method seem to lose its 'sensitivity' at higher levels of disease because the s(E) increases as m increases and therefore the Z statistic decreases as m approaches N.*

6.4.5 Spread of VPV with time.

The number of infected plants in Plot 1 increased rapidly from September to December, 1987 (Table 6.12). This period coincided with a large number of *Aphis gossypii* and another unidentified aphid colonizing *Sonchus oleraceous* and *Emilia sonchifolia* herbaceous weeds found in this plot. The presence of winged aphids within this plot make it highly likely that they helped spread the VPV as detected in the two dimensional or 'quadrat analysis'. These herbaceous weeds were not found in Plot 2 as the grower weeds and clean his plantation on a weekly basis. There were only 2 plants infected in Plot 2 during this period compared to 40 in Plot 1.

Although there was no data on the number of aphids present in Plot 1 during this period, it is known that the number of aphids present in a crop infected with an aphid transmitted virus often correlates to the incidence of infected plants (Thresh, 1986). This type of information had been used, together with weather information, to forecast disease outbreaks from weather patterns that affect aphid population growth (Heathcote, 1986).

6.5 CONCLUSIONS

(i) VPV symptomatic plants are widespread on the main growing islands of Vava'u and Tongatapu. With mostly low incidences in the plantations surveyed, but can be very high in isolated cases.

(ii) The most likely source of VPV in Tonga are infected cuttings introduced when the plantations were established.

(iii) VPV is spread along the Vanilla rows, by growers, during cultural practices like looping, flower initiation and the taking of cuttings.

(iv) Aphid transmission within plantations is highly likely when winged aphid populations are present on herbaceous weeds or grasses, inside the plantation.

....CHAPTER 7....

GENERAL DISCUSSIONS AND CONCLUSIONS

Numerous pathogens have been reported to attack Vanilla in Tonga. Originally, the necrotic spots that follow the diffuse sunken chlorotic streaks on vanilla leaves (Section 5.4.2) was thought to be caused by a fungus. However, the preliminary studies at MAFF Research Station in Tonga and at the University of Florida, Gainesville, suggested that no fungi or bacteria are present, apart from surface contaminants, in leaf and stem tissues with such symptoms.

Further investigations found a potyvirus, referred to as VPV (Vanilla Potyvirus), in this study, but has now been subsequently renamed VNPV (Vanilla Necrosis Potyvirus) in Pearson *et al* (1990) and other later publications, to infect plants that show these symptoms. Two other viruses, Cymbidium Mosaic Virus (CyMV) and Odontoglossum Ringspot Virus (ORSV), were also found in Vanilla plants in Tonga but were not associated with any specific symptoms in this study.

The characterisation of VPV suggest that it is a definite member of the Potyvirus Group. A search of the literature found no reference to this virus on *Vanilla fragrans*. It is regarded as an 'uncited' virus.

Purification of VPV from frozen Vanilla tissue was quite difficult due to irreversible aggregation of the virus particles and therefore virus particles would have been lost in the low speed centrifugation. It was found that increasing the amount of infected leaf tissue partially solved the problem, in that enough potyvirus was extracted for the purpose of developing an antiserum. A second problem is the presence of low molecular weight plant debris in the preparation, together with the virus particles.

A series of sucrose cushions coupled with 2 low speed centrifugations reduced the amount of plant material present as observed under the electron microscope. The presence of immunogenic plant proteins in the virus preparation showed a strong reaction to healthy Vanilla sap when tested using the microprecipitin test. After absorbing out the antibodies to the healthy plant proteins, it was possible to get a clear reaction to VPV infected Vanilla sap.

The antiserum production was successful in producing a working antiserum for use with ELISA, although the reaction was not as strong as the CyMV and ORSV antiserum. However, it was possible to observe a clear distinction between VPV infected Vanilla sap and healthy vanilla sap to be usable for virus testing Vanilla plants. Another antiserum developed by giving the rabbit a 'booster' injection gave an even better differentiation between healthy and VPV infected vanilla plants (Pearson pers. comm.) but it was not used in this study.

The pathogenicity studies on CyMV, ORSV and VPV found VPV to be the most destructive of the 3 viruses on Vanilla plants in Tonga. The symptoms attributed to VPV from preliminary surveys and tests was reproduced through grafts and mechanical inoculations. These symptoms were found comparable to infected plants in the field and those plants also tested positive to VPV using ELISA and ISEM.

The relative incidence of the 3 viruses on *Vanilla fragrans* in Tonga, in this study, were much higher than similar studies on *Vanilla tahitiensis* in french Polynesia (Wisler *et al*, 1987a&b). The relative incidence of CyMV and ORSV, in this study, were also comparable to studies in New Zealand on these same viruses on *Cymbidium* Orchids (Arunasalam, 1987). The greater sensitivity of ELISA, used in the New Zealand and Tongan studies, compared to the Sodium Dodecyl

Sulphate (SDS) - Immunodiffusion test, used in the French Polynesia study, may be the reason for the differences in the number of plants positive to CyMV and ORSV.

Studies of the epidemiology of VPV found this virus to be widespread on the main Vanilla growing islands of Tonga, Vava'u and Tongatapu. The incidence of VPV, based on symptoms only, in the surveyed plantations appear to be low to moderate with very high incidences in isolated cases. Examination of newly planted cuttings found a 100% correlation between symptomatic cuttings and symptomatic shoots. It is concluded that the long spread distance of VPV is apparently due to infected cuttings.

Analysis of the two dimensional distribution pattern in 3 monitored Vanilla plots found the distribution of VPV infected Vanilla plants to be generally clumped. Aphids are the most likely cause of this spread, because (i) aphids have been found to land on Vanilla leaves, in the field (ii) aphids were observed to probe and feed on Vanilla leaves in the laboratory (iii) aphids are known to transmit VPV from infected to healthy plants (iv) aphids were found in large numbers on weeds, in a plantation where the disease spread rapidly. The clumped distribution suggest that the source of infection are nearby infected plants and was highly significant at P=0.001.

Linear distribution of VPV also appear to be clumped in Plot 1, but not in the other 2 plots. This fact suggest that the cultural practices of the growers like looping, flower initiation and the taking of cuttings does spread the viruses along the Vanilla rows.

No serological relationship was found between VPV from *Vanilla fragrans* in Tonga and the potyvirus found in *Vanilla tahitiensis*, in French Polynesia. Potyvirus like particles found in *V.tahitiensis* and *V.fragrans* from the Cook Islands also did not react with the antiserum to the potyvirus found in *V.fragrans* in Tonga. However, a strong reaction was found between a potyvirus found in symptomatic Vanilla leaves from Fiji and the antiserum to VPV. It is known that a large number of Vanilla cuttings in Fiji came from Tonga. The relative symptomatology of these potyviruses appear to be quite different. The potyvirus from Tonga ultimately kills the Vanilla plant whereas the ones from French Polynesia and the Cook Islands appear to be quite mild (Pearson, pers. comm.).

A mild potyvirus may also be present in Tonga as some Vanilla plants with mild symptoms gave weak reactions to ELISA A405nm readings. *Liefting et al (1992) also reported a study of a mild strain of VNPV from Tonga.*

As Vanilla is a very popular crop in the Pacific Islands, because of the high prices in the world market, it is important to put virus and plant protection control measures in place within plantations, villages, islands and also between islands of the Pacific to maintain this excellent source of income for the peoples of the Pacific.

• • • •

RECOMMENDED CONTROL MEASURES FOR VIRUSES OF *VANILLA FRAGRANS* IN THE PACIFIC ISLANDS.

1. Select cuttings for new plantations from vigorous healthy vines and plants. If possible, from plantations that does not have virus infected plants.
2. Remove all plants that show chlorotic, distorted leaves with or without necrosis on stem or leaves, from current plantations. Burn or bury those vines so they do not become sources of infection. For example, if you throw them in a rubbish heap or in the bush, they may still continue to grow for a long time infecting plants nearby when aphid vectors are present. Replace those plants with healthy cuttings.
3. Cuttings that are bought by growers from other villages and islands should be selected carefully to ensure no diseased vines are included.
4. If possible establish new nurseries and plantations with new ELISA tested planting material. Tissue cultured 'virus free' Vanilla plantlets were available from the regional organizations such as the South Pacific Commission (now called the Secretariat for the Pacific Community), Suva, Fiji (Email: spc@ spc.int or phone +679 337 0733) and University of the South Pacific, Alafua Campus, Apia, Samoa (Email: enquiries@samoa.usp.ac.fj or phone 685 21671) during the 1990s when I was working there. They may still be available but are only supplied to member countries on request.

More information is also available from these books by the author available from amazon.com in 'colour' and 'black and white' copies.

1. Plant Protection in the Pacific, Second edition
2. Plant Protection in the Pacific 2
3. Plant Protection in the Pacific 3, Tissue Culture
4. Vanilla Necrosis Potyvirus, Second edition

Or simply search Semisi Pone on the amazon.com website and the list of books will appear on the screen.

References

Additional References;

1. Grisoni, M., Davidson, F., Hyrondelle, C., Farreyrol, K., Caruana, M. L., and Pearson, M. 2004. Nature, incidence, and symptomatology of viruses infecting *Vanilla tahitensis* in French Polynesia. Plant Dis. 88:119-124.

2. Liefting, L., Pearson, M. N., and Pone, S. P. 1992. The isolation and evaluation of two naturally occurring mild strains of Vanilla necrosis potyvirus for control by cross-protection. J. Phytopathol. 136:9-15.

3. Pearson, M.N., Brunt, A.A. and Pone, S.P. (1990) Some Hosts and Properties of a Potyvirus Infecting *Vanilla fragrans* (Orchidaceae) in the Kingdom of Tonga. *Journal of Phytopathology*, Vol. 128. Issue 1, 46-54.

4. van der Grijp, P. Brown Gold. Official expectations and local assessment of the Tongan Vanilla production, Journal de la Société des Océanistes, 104, 1997, 93-103.

5. Y. Y. Wang, D. L. Beck, R. C. Gardner, M. N. Pearson (1993). Nucleotide sequence, serology and symptomatology suggest that vanilla necrosis potyvirus is a strain of watermelon mosaic virus II. Archives of Virology, March 1993, Vol. 129, Issue 1-4, 93-103.

The following are the original references during the writing of the Thesis, 1987-1988. Scanned from the original Thesis.

REFERENCES.

Agrios, G.N. (1978). Plant Pathology, second edition. Academic Press, New
    York, 703 pp.

Alconero, R. (1968). Infection and development of Fusarium oxysporum f.sp.
    vanillae in vanilla roots. Phytopathology 58, 1281-1283.

Alconero, R. (1969a). Mycorrhizal synthesis and pathology of Rhizoctonia
    solani in vanilla orchid roots. Phytopathology 59, 426-430.

Alconero, R. (1969b). Mycorrhizal and parasitic infections of vainilla
    roots. Phyton 26 (1), 17-22.

Alconero, R. and Santiago, A.G. (1969). Mycorrhizal infections of mature
    portions of vanilla roots by Rhizoctonia solani as a
    predisposing factor to infection by Fusarium oxysporum f.sp.
    vanillae. Phytopathology 59, 1521-1524.

Annual Report of the Department of Agriculture, Dominica. (1946).
    Abstract. Review of Applied Mycology 27, 464.

Annual Report of the Department of Agriculture, Dominica. (1949).
    Abstract. Review of Applied Mycology 32, 367.

Ark, P.A. and Thomas, H.E. (1947). Bacterial leaf spot and bud rot of
    orchids caused by Phytomonas cattleyae. Abstract. Review of
    Applied Mycology 26, 60.

Arunasalam, G. (1987). In vitro manipulation of Odontoglossum Ringspot
    Virus and Cymbidium Mosaic Virus in Cymbidium Sw. MSc Thesis,
    University of Auckland, New Zealand.

Ashley, J. (1980). The culture of vanilla in Uganda.
    World Crops 32 (5), 125-129.

Benda, G.T.A and Bennett, C.W. (1964). Effect of curly top virus on
    tobacco seedlings; infection without obvious symptoms.
    Virology 24, 97-101.

Bos, L. (1978). Symptoms of Virus Diseases in Plants. Research Institute
    for Plant Protection, Wagenigen, The Netherlands, 225 pp.

Bos, L. (1975). The application of TMV particles as an internal
    magnification standard for determining virus particle sizes
    in electron microscopes. Netherlands Journal of Plant Pathology
    81, 168-175.

Bouriquet, G. (1932). Madagascar: list of parasites and diseases of
    cultivated plants. Abstract. Review of Applied Mycology 11, 699.

Bouriquet, G. (1934). Les maladies du vanillier a Madagascar [Diseases of
    vanilla in Madagascar]. Abstract. Review of Applied Mycology

13, 57-58.

Bouriquet, G. (1943). Contribution a l'etude des alterations de la vanille preparee (mois issures et mite)[A contribution to the study of alterations in processed vanilla (moulds and a mite)]. Abstract. Review of Applied Mycology 22, 326.

Bouriquet, G. Ed. (1954). Le Vannilier et la Vanille dans le Monde. Paul Lechavalier, Paris, 748 pp.

Bouriquet, G. (1959). Plant diseases and pests in some African teritories. Abstract. Review of Applied Mycology 38, 657.

Brandes, J. and Bercks R. (1965). Gross morphology and serology as a basis for the classification of elongated viruses. Advances in Virus Research 11, 1-22.

Broadbent, L. (1964). Control of Plant Virus Diseases, p 330-364. In Plant Virology. Eds. Corbett M.K. and Sisler H.D. University of Florida Press, 527 pp.

Cibes, H.R. and Childers N.F. (1946). Vanilla, Agronomic Studies pp 36-38. In the Report of the Federal Experiment Station in Puerto Rico, 1946. United States Department of Agriculture.

Cibes, H.R. and Childers N.F. (1949). Abstract on diseases of vanilla in Puerto Rico. Review of Applied Mycology 28, 377.

Ciferri, R. (1926). Sul potere patogenetico dei funghi causanti l 'antracnosi' delle Orchidaceae. [On the pathogenicity of the fungi causing anthracnose of Orchidaceae]. Abstract. Review of Applied Mycology 5, 429.

Ciferri, R. (1928). Mycological and phytopathological notes. Series II. Abstract. Review of Applied Mycology 7, 229.

Clark, M.F. (1981). Immunosorbent Assays in Plant Pathology. Annual Review of Phytopathology 19, 83-106.

Clark, M.F. and Adams, A.N. (1977). Characteristics of the Microplate Method of Enzyme-Linked Immunosorbent Assay for the Detection of Plant viruses. Journal of General Virology 34, 475-483.

Correll, D.S. (1953). Vanilla - Its Botany, History, Cultivation and Economic Import. Economic Botany 7, 291-358.

Daft, G. and Merrick, J. (1973). unpublished. Vanilla survey. Ministry of Agriculture, Fisheries, and Forests, Tonga.

Dequaire, J. (1979) unpublished. A guide to vanilla culture and curing. Ministry of Agriculture, Fisheries, and Forests, Nuku'alofa, Tonga.

Dingley, J.M., Fullerton, R.A., and Mckenzie, E.H.C. (1981). SPEC/UNDP/FAO Survey of Agricultural Pests and Diseases. Volume 2. (Cook Islands, Fiji, Kiribati, Niue, Tonga, Tuvalu, Western Samoa) 485 pp.

Edwardson, J.R. and Zettler, F.W. (1986). Odontoglossum Ringspot Virus.
pp 233-247. In The Plant Viruses. Eds. M.H.V. van Regenmortel
and H. Fraenkel-Conrat. Plenum Press, New York, 424 pp.

Elliot, J.M. (1977). Some methods for the statistical analysis of benthic
invertebrates. Second edition. Freshwater Biological
Association Scientific Publications No. 25. United Kingdom.
160 pp.

Engvall, E. and Perlmann, P. (1971). Enzyme-linked immunosorbent assay
(ELISA). Quantitative assay of immunoglobulin G.
Immunochemistry 8, 871-874.

Engvall, E. and Perlmann, P. (1972). Enzyme-linked immunosorbent assay,
ELISA III. Quantitation of specific antibodies by enzyme
labelled anti-immunoglobulin in antigen coated tubes.
Journal of Immunology 109, 129-135.

Fa'anunu, H.'O. (1985) unpublished. Annual Report of the Vanilla Project
Manager, Ministry of Agriculture, Fisheries and Forests (MAFF),
Nuku'alofa, Tonga.

Fiester, D.R. and Elliot, R.C. (1966). Recomendaciones para la explotacion
de vainilla en Guatemala [Recommendations for the cultivation
of vanilla in Guatemala]. Abstract. Review of Applied Mycology
45, 272.

Francki, R.I.B. (1966). Isolation, purification, and some properties of two
viruses from cultivated Cymbidium orchids. Australian Journal
of Biological Science 19, 555-564.

Francki, R.I.B. (1970). Cymbidium Mosaic Virus. CMI/AAB Description of
plant viruses, Number 27. 3 pp.

Francki, R.I.B. (1972). Purification of Viruses. In Principles and
Techniques in Plant Virology. Eds. Kado, C.I. and Agrawal, H.O.
688 pp.

Fruit and Tropical Products. Commonwealth Secretariat. London. December,
1987.

Fulton, R.W. (1964). Transmission of viruses by grafting, dodder, seed, and
mechanical inoculation pp 39-67. In Plant Virology. Eds.
M.K.Corbett and H.D. Sisler. University of Florida Press,
525 pp.

Gibbs, A. (1976). Viruses and Plasmodesmata, pp 149-164. In Intercellular
Communication in Plants: Studies on Plasmodesmata.
Eds. B.E.S. Gunning and A.W. Robards. Springer-Verlag, Berlin,
387 pp.

Gibbs, A. and Harrison B. (1976). Plant Virology, The Principles.
Edward Arnold, London, 292 pp.

Gibbs, A. (1983). A simple convolution method for describing or comparing
     the distributions of virus affected plants in a plant
     community, pp 39-50. In Plant Virus Epidemiology. Eds. R.T. Plumb
     and J.M. Thresh. Blackwell Scientific Publications, Oxford,
     pp 377.

Hamilton, R.I., Edwardson, J.R., Francki, R.I.B., Hsu, H.T., Hull, R.,
     Koenig, R., and Milne, R.G. (1981). Guidelines for the
     identification and characterisation of plant viruses. Journal
     of General Virology 54, 223-241.

Heathcote, G.D. (1986). Virus yellows of sugar beet. pp 399-417.
     In Plant Virus Epidemics, Monitoring, Modelling, and
     Predicting Outbreaks. Eds. Mclean, G.D., Garret, R.G.,
     and Ruesink, W.G. Academic Press, Sydney. 550 pp.

Hill, S.A. (1984). Methods in Plant Virology. Blackwell Scientific
     Publications, London. pp 167.

Hine, R.B., Osborne, W.E. and Dennis, R.E. (1970). Elevation and temperature
     effects on severity of maize dwarf mosaic virus in sorghum in
     Arizona. Plant Disease Reporter 54, 1064-1068.

Hollings, M. and Brunt, A. (1981a). Potyvirus Group. CMI/AAB Descriptions
     of Plant Viruses. No.245. 7pp.

Hollings, M. and Brunt, A. (1981b). Potyviruses. In Handbook of Plant
     Virus Infections and Comparative Diagnosis. pp 731-807.
     Ed. E. Kurstak. Elsevier/North-Holland Biomedical Press,
     943 pp.

Horner, M.B. (1986). Investigations and Elimination of a "mottle" virus
     from Polianthes tuberosa L. MSc Thesis, University of Auckland,
     New Zealand.

Jensen, J.H. (1940). Plant disease investigations. Abstract. Review of
     Applied Mycology 19, 201.

Jensen, D.D. (1951). Mosaic or Black Streak Disease of Cymbidium Orchids.
     Phytopathology 41, 401-413.

Jensen, D.D. and Gold, A.H. (1951). A virus ringspot of Odontoglossum
     orchids. Symptoms, transmission, and electron microscopy.
     Phytopathology 41, 648-653.

Jensen, D.D. and Gold, A.H. (1955). Hosts, Transmission, and Electron
     Microscopy of Cymbidium Mosaic Virus with special reference to
     Cattleya Leaf Necrosis. Phytopathology 45, 327-334.

Johnstone, G.R., Dufus, J.E., Munro, D., and Ashby, T.W. (1982).
     Purification of a Tasmanian isolate of subterranean clover red
     leaf virus, and its serological interactions with a New Zealand
     isolate and other luteoviruses. Australian Journal of
     Agricultural Research 33, 697-703.

Kado, C.I. and Jensen, D.D. (1964). Cymbidium Mosaic Virus in Phalaenopsis. Phytopathology 54, 975-977.

Kopp, A. (1930). Les maladies des plantes a la Reunion [Plant Diseases in Reunion]. Abstract. Review of Applied Mycology 9, 766-767.

Leakey, C.L.A. (1970). The role of fungi and cultural factors in the etiology of root rot of vanilla (Vanilla fragrans (Salisb.) Ames) in Uganda. East African Agricultural and Forestry Journal. April 1970, 207-209.

Leseman, D.E. and Koenig, R. (1985). Identification of bean yellow mosaic in Masdevallia. Acta Horticulturae 164, 347-354.

Leseman, D.E. and Vetten, J.J. (1985). The occurrence of tobacco rattle and turnip mosaic viruses in Orchid spp. and of an unidentified potyvirus in Cypripedium calceolus. Acta Horticulturae 164, 45-54.

Lionnet, J.F.G. (1958). Vanilla cultivation in the Seychelles Islands I. World Crops 10, 441-444.

Madden, L.V. and Campbell, C.L. (1986). Descriptions of virus disease epidemics in time and space. pp 273-293. In Plant Virus Epidemics, Monitoring, Modelling, and Predicting outbreaks. Eds. Mclean, G.D., Garret, R.G., and Ruesink, W.G. Academic Press, Sydney. pp 550.

Madden, L.V., Louie, R., Abt, J.J., and Knoke, J.J. (1982). Evaluation of tests for randomness of infected plants. Phytopathology 72, 195-198.

Martin, F.W. (1963). Chromosome number and behaviour in a vanilla hybrid and several vanilla species. Bulletin of the Torrey Botanical Club 90 (6), 416-417.

Matthews, R.E.F. (1970). Plant Virology. Academic Press, New York. 778 pp.

Matthews, R.E.F. (1981). Plant Virology, second edition. Academic Press, New York, 778 pp.

Matthews, R.E.F. (1979). The classification and Nomenclature of viruses, summary of results of the meeting of the International Committee of Taxonomy of Viruses in The Hague, September 1978. Intervirology 11, 133-157.

Matthews, R.E.F. (1982). Classification and Nomenclature of Viruses. Fourth Report of the International Committee on Taxonomy of Viruses. Intervirology 17, 1-200.

Maublanc, A. and Barat, H. (1927). Une maladie nouvelle de la vanille [A new disease of vanilla]. Abstract. Review of Applied Mycology 6, 599.

Moreau, C. and Moreau, M. (1962). Quelques alterations des gousses de vanilla a Madagascar [Some infections of vanilla pods in Madagascar]. Abstract. Review of Applied Mycology 41, 165.

Mossop, D.W. and Fry, P.R. (1984). SPEC/UNDP/FAO Survey of Agricultural Pests and Diseases: Records of viruses pathogenic on plants in the Cook Islands, Fiji, Kiribati, Niue, Tonga, and Western Samoa, Volume 7. 9 pp.

Muzik, T.J. (1958). Role of parenchyma cells in graft union in vanilla Orchids. Science 127 (3289), 82.

Noordam, D. (1973). Identification of Plant viruses. Centre for Agricultural Publishing and Documentation, Wagenigen. 207pp.

Paul, H.L. (1975). Odontoglossum Ringspot Virus. CMI/AAB Descriptions of Plants viruses. No. 155. 4 pp.

Pearson, M.N. and Cole, J.S. (1986). The effects of Cymbidium Mosaic virus and Odontoglossum Ringspot Virus on the growth of Cymbidium orchids. Journal of Phytopathology 117, 193-197.

Pearson, M.N. and Pone, S.P. (1988) Viruses of vanilla in the Kingdom of Tonga. Australian Plant Pathology 17 (3), 59-60.

Pearson, M.N.; Pone, S.P., and Brunt, A.A. (unpublished.) Some hosts and properties of a potyvirus from Vanilla fragrans in the Kingdom of Tonga.

Petch, T. and Ragunathan, C. (1927). Abstract. The fungi associated with disease in vanilla. Review of Applied Mycology 6, 694-695.

Plenet, R. (1958). Note sur la maladie noive du vanillier a la Reunion. [A note on black disease of vanilla in Reunion]. Abstract. Review of Applied Mycology 37, 365.

Pone, S.P, and Nai, P. V. (1985). unpublished. Vanilla Isolations - Preliminary Report. Research Division, MAFF, Nuku'alofa, Tonga.

Project Performance Audit Report. unpublished. Tonga Development Bank, May 1988.

Purseglove, J.W., Brown, E.G., Green, C.L., and Robbins, S.R.J. (1981). Vanilla. pp 644-735. In Spices, Volume 2. Longmans.

Ragunathan C. (1924). Soft rot of Vanilla planifolia. Abstract. Review of Applied Mycology 3, 553-554.

Report of the Phytopathology Division, Madagascar. (1963). Abstract. Review of Applied Mycology 42, 126.

Richter, W. (1965). The orchid world. Studio Vistas, London. 291 pp.

Roberts, D.A. (1964). Local lesion assay for plant viruses. pp 194-220. In Plant Virology. Eds. M.K. Corbett and H.D. Sisler. University of Florida Press, 527pp.

Ross, A.F. (1964). Identification of plant viruses, pp 68-92.

147

In _Plant Virology_. Eds. Corbett, M.K. and Sisler, H.D.
University of Florida Press, Gainesville, 527 pp.

Ross, J.P. (1970). Effect of temperature on mottling of soybean seed
caused by soybean mosaic virus. _Phytopathology_ 60, 1798-1800.

Rybicki, E.P. and von Wechmar M.B. (1985). Serology and Immunochemistry.
pp 207-244. In _The Plant Viruses, polyhedral virions and
tripartite genomes_. Ed. R.I.B. Francki, Plenum Press, New York.
309 pp.

Smith, K.M. (1974). _Plant viruses_, fifth edition. Chapman and Hall,
London. 211 pp.

Smith, E.F. and Godfrey, G.H. (1922). Bacterial wilt of castor bean
(Ricinus communis L.). Abstract. _Review of Applied Mycology_
1, 201-202.

Stehle, H. (1954). Le vanillier et sa culture III; Maladies, insectes, et
affections diverses du vanillier [The vanilla plant and its
cultivation III; Diseases, insects, and various disorders of
the vanilla plant]. Abstract. _Review of Applied Mycology_
33, 183.

Stehle, H. (1955). Quelques notes sur la botanique et l'Archipeldes
Caraibes [Some notes on the botany and plant ecology of the
Carribean Archipelago]. Abstract. _Review of Applied Mycology_
34, 517-518.

Stier, H. (1984). unpublished. Nematodes on vanilla and their relationship
to vine dying. unpublished. Research Division, MAFF, Tonga.

Theis, T. and Jimenez, F.A. (1957). A vanilla hybrid resistant to
Fusarium root rot. _Phytopathology_ 47, 579-581.

Theodose, R. (1973). Traditional methods of vanilla preparaton and their
improvement. _Tropical Science_ 15 (1), 47-57.

Thresh, J.M. (1986). Plant virus disease forecasting, pp 359-386.
In _Plant Virus Epidemics, Monitoring, Modelling, and Predicting
Outbreaks_. Eds. Mclean, G.D., Garret, R.G., and Ruesink, W.G.
Academic Press, Sydney, 550 pp.

Tiollier, V. (1980). unpublished. _Vanilla cultivation in Tonga_. Technical
Bulletin No.1, Research Division, MAFF, Nuku'alofa, Tonga.

Torrance, L. and Jones, R.A.C. (1981). Recent developments in serological
methods suited for use in routine testing for plant viruses.
Plant Pathology 30, 1-24.

Tsao, P.H. (1985). unpublished. _Phytophora_ blight and root rot of vanilla
in French Polynesia: occurence and severity of the disease and
identity of causal species. November, 1985. Report to the French
Polynesian Ministry of Agriculture.

Tucker, C.M. (1924). Report of the Plant Pathologist (Federal Experiment
Station, Puerto Rico). Abstract. _Review of Applied Mycology_

4, 210-211.

Tucker, C.M. (1925). Report of the Plant Pathologist (Federal Experiment Station, Puerto Rico). Abstract. Review of Applied Mycology 5, 283-284.

Tucker, C.M. (1927). Report of the Plant Pathologists (Federal Experiment Station, Puerto Rico). Abstract. Review of Applied Mycology 6, 559-602.

Turconi, M. (1924). Note di patologia vegetale. 1. Un'infezione di Botrytis cinerea Pers. in giovani frutti di vaniglia (Vanilla planifolia Andr.)[Notes on Plant Pathology. 1. An infection by Botrytis cinerea Pers. of young vanilla fruits (Vanilla planifolia Andr.) Abstract. Review of Applied Mycology 3, 427.

van der Plank (1960). Analysis of epidemics, pp 229-289. In Plant Pathology, Volume III. Eds. Horsfall, J.G. and Dimond, A.E., Academic Press, New york, 675 pp.

van Regenmortel, M.H.V. (1966). Plant Virus Serology. Advances in Virus Research 12, 207-271.

van Regenmortel, M.V.H. (1982). Serology and Immunochemistry of Plant Viruses. Academic Press, London. 302 pp.

van Velsen, R.J. (1979). unpublished. Final Report of UNDP/FAO survey of plant viruses in the Kingdom of Tonga, MAFF, Tonga.

Verwoerd, L. and DuPleiss, S.J. (1934). Description of some new species of South African fungi and of species not previously reported from South Africa. Abstract. Review of Applied Mycology 13, 183-184.

Voller, A., Bartlett, A., Bidwell, D.E., Clark, M.F., and Adams, A.N., (1976). The Detection of Viruses by Enzyme-Linked Immunosorbent Assay (ELISA). Journal of General Virology 33, 165-167.

Walkey, D.G.A. (1985). Applied Plant Virology. Heinemann, London, pp 329.

Wisler, G.C., Zettler, F.W., and Mu, L. (1987a). Viruses affecting vanilla in French Polynesia. American Orchid Society Bulletin 56 (4), 1987.

Wisler, G.C., Zettler, F.W., and Mu, L. (1987b). Virus infections of Vanilla and other orchids in French Polynesia. Plant Disease 71 (12), 1125-1129.

Wisler, G.C., Zettler, F.W., and Pursifull, D.E. (1982). A serodiagnostic technique for detecting Cymbidium Mosaic virus and Odontoglossum Ringspot Viruses. Phytopathology 72 (7), 835-837.

Wisler, G.C., Zettler, F.W., and Sheehah, T.J. (1979). Relative incidence of Cymbidium Mosaic Virus and Odontoglossum Ringspot Viruses in several genera of wild and cultivated orchids. Proceedings of the Florida State Horticultural Society 92, 339-340.

Wrigley, N.G. (1968). The lattice spacing of crystalline catalase as an internal standard of length in electron microscopy. <u>Journal of Ultrastructure Research</u> 24, 457–461.

Young, B.R. and Blundell, W.L. (1979). Orchid viruses in New Zealand. <u>New Zealand Orchid Review</u>, December 1979, 3–5.

Zettler, F.W., Hennen, G.R., Bodranuk, W.H. Jr, Clifford, H.T., and Sheehah, T.J. (1978). <u>Plant Disease Reporter</u> 62 (11), 949–952.

PERSONAL COMMUNICATIONS.

Pearson, M.N. – Botany Department, University of Auckland, N.Z.

Stechmann, D.H. – Am Buhlweiher 1, D-8581 Donndorf, Federal Republic of

Germany.

van Velsen, R.J. – Department of Agriculture and Fisheries, Adelaide,

South Australia, Australia.

APPENDIX 1. CHEMICALS USED IN ELISA.

Phosphate buffered saline (PBS)

8.0 g NaCl
0.2 g KH2P04
1.15 g Na2HP04
0.2 g KC1
Make up to 1 litre with deionised water, pH 7.4

PBS—tween

PBS + 0.5 ml   'Tween 20' per litre

Coating buffer

4.29 g Na2C03.10H20
2.93 g NaHCOg
Make up 1 litre with deionised water, pH 9.6

Substrate buffer

97 ml Diethanolamine
800 ml H20
Adjust to pH 9.8 with HC1 and make up to 1 litre with deionised water.

Phosphate substrate

3x 5 mg phosphate substrate (Disodium p-Nitrophenyl- Phosphate, Sigma104 No 104-105) tablets dissolved in 25 ml of substrate buffer.

APPENDIX 2. A405 VALUES OF VPV, CyMV AND ORSV SAMPLES AT DIFFERENT COATING GAMMA GLOBULIN, ENZYM CONJUGATE AND SAP EXTRACT.

VPV

			10 ul			1 ul			0.1 ul		
Test samples	I	1:1	0.246	0.208	0.266	0.174	0.047	0.057	0.200	0.031	0.022
		1:10	0.216	0.259	0.022	0.156	0.052	0.049	0.096	0.025	0.035
		1:100	0.319	0.205	0.050	0.251	0.059	0.119	0.303	0.061	0.011
	II	1:1	0.339	0.177	0.044	0.217	0.098	0.117	0.262	0.212	0.035
		1:10	0.256	0.148	0.053	0.146	0.166	0.088	0.109	0.153	0.113
controls		1:1	0.039	0.021	0.009	0.020	0.010	0.013	0.022	0.009	0.009
		1:10	0.035	0.018	0.006	0.013	0.007	0.011	0.011	0.012	0.006
PBS-tween			0.019	0.019	0.008	0.014	0.011	0.016	0.018	0.036	0.010
Enzyme conjugate			1:400	1:800	1:1600	1:400	1:800	1:1600	1:400	1:800	1:1600

CyMV

			10 ul			1 ul			0.1 ul		
Test samples	I	1:1	1.298	0.617	0.285	0.684	0.370	0.145	0.335	0.132	0.054
		1:10	1.038	0.526	0.252	0.776	0.377	0.146	0.311	0.144	0.062
	II	1:1	0.106	0.051	0.020	0.071	0.032	0.010	0.051	0.032	0.007
		1:10	0.135	0.045	0.021	0.074	0.030	0.018	0.065	0.025	0.014
controls		1:1	0.092	0.033	0.015	0.112	0.033	0.013	0.052	0.027	0.034
		1:10	0.111	0.045	0.017	0.092	0.029	0.013	0.032	0.013	0.005
PBS-tween			0.059	0.032	0.008	0.032	0.029	0.011	0.040	0.013	0.004
Enzyme conjugate			1:400	1:800	1:1600	1:400	1:800	1:1600	1:400	1:800	1:1600

ORSV

			10 ul			1 ul			0.1 ul		
Test samples	I	1:10	1.571	0.652	0.283	1.077	0.413	0.198	0.352	0.176	0.063
		1:100	1.587	0.626	0.275	1.135	0.454	0.206	0.559	0.254	0.094
	II	1:1	0.266	0.119	0.037	0.173	0.076	0.041	0.157	0.056	0.023
		1:10	0.120	0.057	0.007	0.093	0.017	0.201	0.118	0.044	0.010
		1:100	0.175	0.042	0.002	0.123	0.009	0.050	0.232	0.230	0.009
controls		1:1	0.109	0.038	0.000	0.275	0.121	0.034	0.139	0.071	0.011
		1:10	0.064	0.050	-0.004	0.122	0.181	0.031	0.106	0.050	0.021
PBS-tween			0.081	0.029	0.007	0.079	0.015	0.003	0.100	0.030	0.040
Enzyme conjugate			1:400	1:800	1:1600	1:400	1:800	1:1600	1:400	1:800	1:1600

I - high virus concentration test sample.

II - low virus concentration test sample.

APPENDIX 3. Locations of plantings surveyed for VPV symptomatic plants on Tongatapu (T12 not marked).

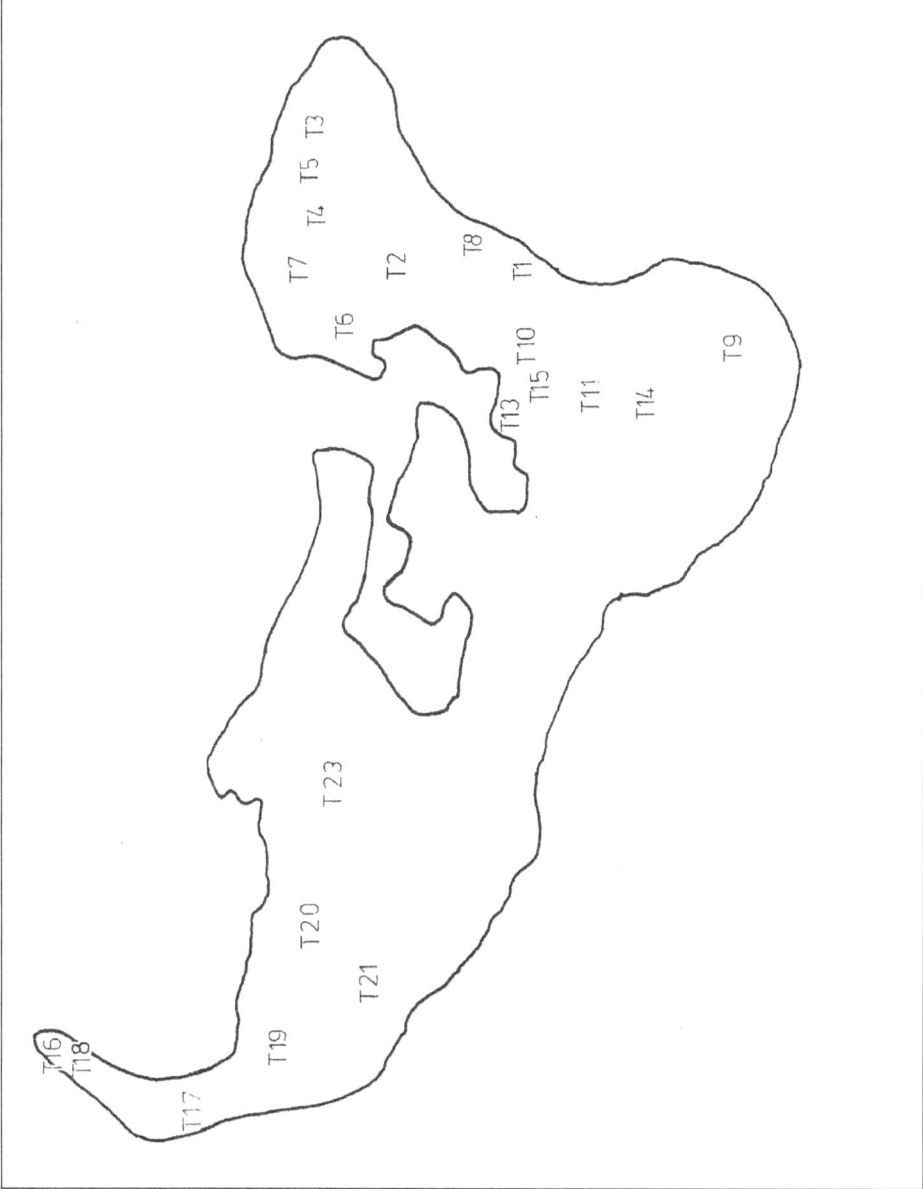

APPENDIX 4. Locations of plantings surveyed for VPV symptomatic plants on the Vava'u main island (V17, V18, V29, and V30 not marked)

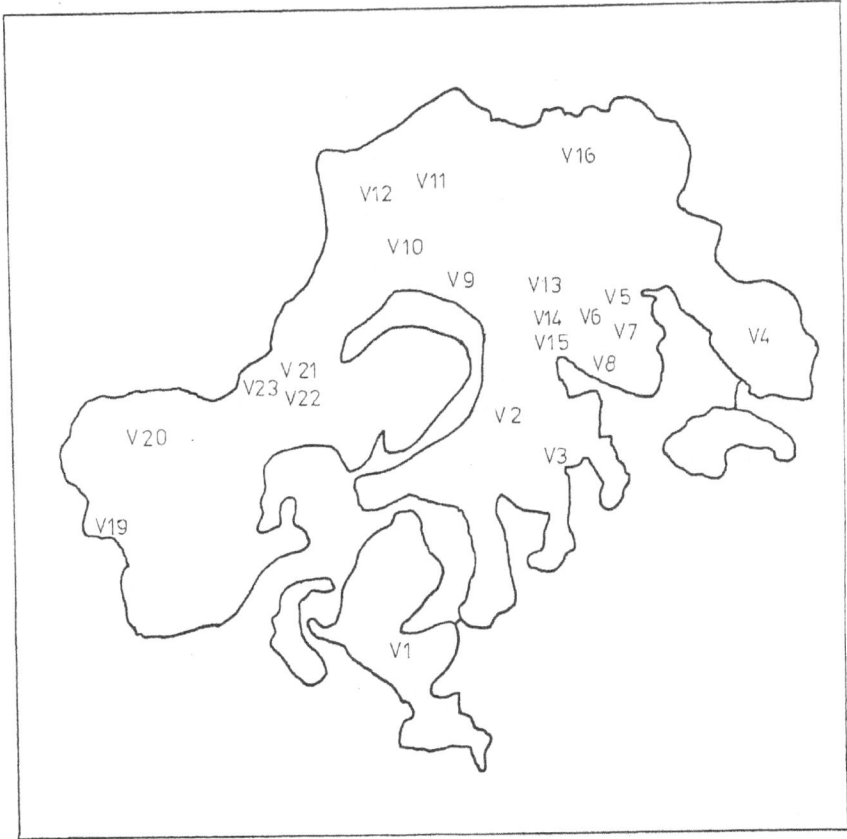

APPENDIX 5. List of plantations surveyed and the respective owners.

APPENDIX 5. List of plantations surveyed and the respective owners.

Tongatapu   Vava'u

Plantation	Owner	Plantation	Owner
T1	Kailahi Ngungutau	V1	Selemaia Te'ekiu
T2	Petelo Sifa	V2	Filipe Hu'ahulu
T3	Kisi Hafoka	V3	Sateki Lea
T4	Sione Pakofe	V4	Viliami Hopo
T5	Filipe Maile	V5	Ve'emolo Matangi
T6	Uelesili Tu'i'onetoa	V6	Simione Tupou
T7	Mo'unga Manumu'a	V7	Finepolo Tupou
T8	Ma'ulupe Ngungutau	V8	Lea'aevai Halaifonua
T9	Heneli Raas	V9	S. Laulile Lavulavu
T10	Prince Mailefihi Tuku'aho	V10	Tevita Paea
T11	Lafaele Matakaiongo	V11	Lupe Takeifanga
T12	Sione Manufekai	V12	Sakalaia Moala
T13	Soakimi Ha'ungatau	V13	Uasi Katoa
T14	Pasiaka 'Enosi	V14	Fe'iloaki Niusini
T15	Semisi Kutu	V15	'Ikani Latu
T16	Fangupo Mani	V16	Taniela Falekaono
T17	Tonga Luke	V17	His Majesty King – Taufa'ahau Tupou IV
T18	'Ofa Hauati		
T19	Leini Vakalahi	V18	Ha'alefo
T20	Mafana Manoa	V19	Viliami Foketi
T21	Sivoki Tu'ikolongahau	V20	Angahiki Loseli
T23	Lopeti Uhi	V21	Heneli Tu'inukuafe
		V22	Sione Funaki
		V23	Tevita Kolope
		V29	Lisiate Fifita
		V30	Filipe Hu'ahulu

Plantations missing from this list were not included in the original list, but had been inspected during the survey. As they had not been randomly selected it was decided to leave them out. These include T22, V24, V25, V26, V27, and V28.

KO E MAHAKI VAILASI 'O E VANILA'

VIRUS DISEASES OF VANILLA IN TONGA

Fika 1. Ko e fōtunga 'o e vanila 'oku ma'u 'e he vailasi' (fakatokanga'i e mate 'a e ngaahi va'a vanila').
Fig. 1. Disease symptoms on virus infected vanilla (note dying, leafless vines to left of photograph).

ABOUT THE AUTHOR...

Semisi Pule a.k.a. Semisi Pule Pone was born and raised in the Kingdom of Tonga, South Pacific. He attended the local Longolongo Primary School from 1967 to 1973 and successfully passed the entrance examination to Tonga High School. He started at Tonga High School in1974 and was successful in both the New Zealand School Certificate Exam (1978) as well as the University Entrance Exam (1979), passing 5 subjects in both. He continued his studies at Mt Albert Grammar School in 1980 and successfully completed the requirements to enter Auckland University in 1981. In May, 1985 he graduated with a Bachelor of Science and returned to Tonga in June, 1985.

He was offered a job as an Agriculture Officer/Plant Pathologist and began working for the Ministry of Agriculture Fisheries and Forests in June 1985. Three of his most important projects were **1. Vanilla viruses 2. Kava virus dieback** and **3. Squash/Pumpkin viruses**. He continued on to do a Master of Science (Honours) degree, at Auckland University, in 1987 graduating in 1989. During the Squash boom years he helped MAFF advise the Tongan squash growers on how to control the Zucchini Yellow Mosaic Virus which destroyed many plantations. Tonga was able to export 22, 000 tonnes of squash to Japan in 1991 despite the virus problem. The results of his work in the Pacific are published in his series of books PLANT PROTECTION IN THE PACIFIC and several advisory ebooks available from amazon.com.

In April 1992, he joined the Institute for Research, Extension and Training in Agriculture (IRETA) at the University of the South Pacific Agriculture Campus, Apia, Samoa. He was responsible for maintenance and dissemination of elite and popular Pacific Island crops stored in Tissue Culture. Popular crops such as Sweet Potato,

Vanilla, Yam, Taro, Cassava and Bananas were sent to USP member countries, on request, for research and hurricane recovery programmes . He also carried out research on mass multiplication of Vanilla and Sweet Potato as well as the effect of salinity on Pacific Island crops as sea levels rise and contaminate underground water lens in small atolls. This work is also published in PLANT PROTECTION IN THE PACIFIC 2, Tissue Culture, also available from amazon.com.

He was offered the position of Plant Protection Advisor with the South Pacific Commission Agriculture Programme based in Suva, Fiji in May 1996. During his time with SPC (May,1993-May,1996) he was responsible for more than $NZ30 million in projects as Co-ordinator of the Plant Protection Service. He was also the Manager of the $NZ 5 million SPC/EU Pacific Plant Protection Project and Chief Executive of the newly established Pacific Plant Protection Organization. Much of this work is published in his book CONFESSIONS OF A SENIOR PLANT VIROLOGIST AND PATHOLOGIST also available from amazon.com.

He was selected to represent the Pacific Islands on the Committee of Experts on Phytosanitary Measures which met at the United Nations, Food and Agriculture Organization in Rome every two years and a member of the Regional Plant Protection Organization Technical Consultation which also meet at UN, FAO Rome every other year.

In June 1996, he migrated with his family to New Zealand. He was involved in many industries for 15 years, then he started writing   in 2011 and has published more than 200 books and ebooks in amazon.com and blurb.com. He also operate a small contracts business.

www.ingramcontent.com/pod-product-compliance
Lightning Source LLC
Chambersburg PA
CBHW021928190326
41519CB00009B/944